ENVIRONMENTAL SCIENCE, ENGINEERING AND TECHNOLOGY

AIR QUALITY AND THE NATURAL GAS INDUSTRY

ISSUES, CONSIDERATIONS, AND REGULATION

ENVIRONMENTAL SCIENCE, ENGINEERING AND TECHNOLOGY

Additional books in this series can be found on Nova's website under the Series tab.

Additional E-books in this series can be found on Nova's website under the E-book tab.

ENVIRONMENTAL HEALTH - PHYSICAL, CHEMICAL AND BIOLOGICAL FACTORS

Additional books in this series can be found on Nova's website under the Series tab.

Additional E-books in this series can be found on Nova's website under the E-book tab.

ENVIRONMENTAL SCIENCE, ENGINEERING AND TECHNOLOGY

AIR QUALITY AND THE NATURAL GAS INDUSTRY

ISSUES, CONSIDERATIONS, AND REGULATION

JAMIE D. ZALESKI
EDITOR

nova publishers
New York

For permission to use material from this book please contact us:
Telephone 631-231-7269; Fax 631-231-8175
Web Site: http://www.novapublishers.com

Library of Congress Cataloging-in-Publication Data

ISBN: 978-1-62618-476-3

Published by Nova Science Publishers, Inc. † New York

CONTENTS

PREFACE

This book focuses on natural gas, both as a saleable product and as a potential air pollutant. After briefly outlining the production, processing, transmission, and distribution phases of the natural gas industry, the book characterizes the types and sources of pollutants, and explores the role of the federal government in regulating these emissions, including the provisions in the Clean Air Act and the regulatory activities of the EPA.

Chapter 1 - Congressional interest in U.S. energy policy has focused in part on ways through which the United States could secure more economical and reliable fossil fuel resources both domestically and internationally. Recent expansion in natural gas production, primarily as a result of new or improved technologies (e.g., hydraulic fracturing, directional drilling) used on unconventional resources (e.g., shale gas, coal-bed methane), has made natural gas an increasingly significant component in the U.S. energy supply. This expansion, however, has prompted renewed questions about the industry's potential impacts on human health and the environment, including impacts on air quality. But unlike the debate over groundwater contamination or induced seismicity—where questions remain as to whether or not production activities contribute significantly to these impacts—there is little question that natural gas systems emit air pollutants.

Chapter 2 – This report summarizes the Clean Air Act and its major regulatory requirements.

The principal statute addressing air quality concerns, the Clean Air Act was first enacted in 1955, with major revisions in 1970, 1977, and 1990. The Act requires EPA to set health-based standards for ambient air quality, sets deadlines for the achievement of those standards by state and local governments, and requires EPA to set national emission standards for large or ubiquitous sources of air pollution, including motor vehicles, power plants, and other industrial sources. In addition, the Act mandates emission controls for sources of 187 hazardous air pollutants, establishes a cap-and-trade program to limit acid rain, requires the prevention of significant deterioration of air quality in areas with clean air, requires a program to restore visibility impaired by regional haze in national parks and wilderness areas, and implements the Montreal Protocol to phase out most ozone-depleting chemicals.

This report describes the Act's major provisions and provides tables listing all major amendments, with the year of enactment and Public Law number, and cross-referencing sections of the Act with the major U.S. Code sections of the codified statute.

In: Air Quality and the Natural Gas Industry
Editor: Jamie D. Zaleski
ISBN: 978-1-62618-476-3

Chapter 1

AIR QUALITY ISSUES IN NATURAL GAS SYSTEMS*

Richard K. Lattanzio

SUMMARY

Natural Gas Systems and Air Pollution

Congressional interest in U.S. energy policy has focused in part on ways through which the United States could secure more economical and reliable fossil fuel resources both domestically and internationally. Recent expansion in natural gas production, primarily as a result of new or improved technologies (e.g., hydraulic fracturing, directional drilling) used on unconventional resources (e.g., shale gas, coal-bed methane), has made natural gas an increasingly significant component in the U.S. energy supply. This expansion, however, has prompted renewed questions about the industry's potential impacts on human health and the environment, including impacts on air quality. But unlike the debate over groundwater contamination or induced seismicity—where questions remain as to whether or not production activities contribute significantly to these impacts—there is little question that natural gas systems emit air pollutants. The concerns, instead, are:

---

* This is an edited, reformatted and augmented version of Congressional Research Service, Publication No. R42833, dated November 26, 2012.

- which pollutants?
- how much of each pollutant?
- from which sources?
- what are the impacts of the emissions?
- how much is the cost of abatement? and,
- what is the role of the federal government?

Air pollutants are released in natural gas systems through the leaking, venting, and combustion of raw natural gas; the combustion of other fossil fuel resources; and the discharge of particulate matter during associated operations. Pollutants include, most prominently, methane and volatile organic compounds—of which the natural gas industry is one of the highest emitting industrial sectors in the United States—as well as nitrogen oxides, sulfur dioxide, particulate matter, and various forms of hazardous air toxics. Emissions sources include pad, road, and pipeline construction; drilling, completion, and flowback activities; and gas processing and transmission equipment such as controllers, compressors, dehydrators, pipelines, and storage vessels.

EPA's New Air Standards

The U.S. Environmental Protection Agency (EPA), in response to a consent decree issued by the U.S. Court of Appeals, D.C. Circuit, promulgated air standards for new and modified source categories in crude oil and natural gas systems on August 16, 2012. These standards—effective October 15, 2012—revised existing rules and promulgated new ones to regulate emissions of volatile organic compounds (VOCs), sulfur dioxide, and hazardous air pollutants (HAPs) from many source categories that had never before been covered by federal oversight. Since natural gas is both the product and a pollutant of the industry, compliance to the standards has the potential to translate into economic benefits, as producers can offset pollution abatement costs with the value of products recovered and sold at market. With this expectation, EPA reported the annual environmental and the economic benefits of the standards to be VOCs reductions of 190,000 tons, HAPs reductions of 12,000 tons, methane reductions of 1.0 million tons, and a net cost savings of $11 million to $19 million. Industry and other stakeholders have disputed these figures as both too high and too low. Further, the expansion of both industry production and government regulation of natural gas has sparked debate on a number of critical issues, including:

- establishing comprehensive emissions data,
- determining the proper coverage of pollutants and sources,
- understanding the human health and environmental impacts of emissions,
- estimating total compliance costs, and
- managing the role of local, state, and federal governments.

Scope and Purpose of This Report

The focus of this report is on natural gas—both as a saleable product and as a potential air pollutant. After briefly outlining the production, processing, transmission, and distribution phases of the natural gas industry, the report characterizes the types and sources of pollutants. It then turns to the role of the federal government in regulating these emissions, including the provisions in the Clean Air Act and the regulatory activities of the EPA. The report summarizes both the prior New Source Performance Standards and National Emission Standards for Hazardous Air Pollutants for the Oil and Natural Gas Sector, and those promulgated on August 16, 2012. It concludes with an extended discussion of the aforementioned critical issues.

BACKGROUND

Congressional interest in U.S. energy policy has focused in part on ways through which the United States could secure more economical and reliable fossil fuel resources both domestically and internationally. Recent expansion in natural gas production, primarily as a result of new or improved technologies (e.g., hydraulic fracturing) used on unconventional resources (e.g., shale, tight sands, and coal-bed methane), has made natural gas an increasingly significant component in the U.S. energy supply.[1] While the practice of hydraulic fracturing is not new, relatively recent innovations have incorporated processes such as directional drilling, high-volume slick-water injection, and multistage fractures to get to previously unrecoverable or uneconomical resources. As a result, the United States has again become the largest producer of natural gas in the world. The U.S. Energy Information

Administration (EIA) projects unconventional gas activity to grow fourfold from 2009 to 2035, and forecasts that it will make up 47% of total U.S. production by 2035.[2] The increased production has served to reduce natural gas prices, resulting in the displacement of coal-fired electric generating units, among other effects. Some analysts believe that the exploitation of new unconventional resources has the potential to reshape energy policy at national and international levels—altering geopolitics and energy security, recasting the economics of energy technology investment decisions, and shifting trends in greenhouse gas (GHG) emissions.[3]

Many in both the public and private sector have advocated for the increased production and use of natural gas because the resource is domestically available, economically recoverable, and considered a potential "bridge" fuel to a less polluting and lower GHG-intensive economy.[4] Natural gas is cleaner burning than its hydrocarbon rivals, emitting, on average, about half as much carbon dioxide as coal and one-quarter less than oil when consumed in a typical electric utility plant.[5] Further, natural gas combustion emits no mercury—a persistent, bioaccumulative neurotoxin—and less particulate matter, sulfur dioxide, and nitrogen oxides, on average, than either coal or oil. For these reasons, pollution control measures in natural gas systems have traditionally received less attention relative to those in other hydrocarbon industries. But with the development of new and improved technologies, as well as the gas drilling boom they have engendered, a different set of concerns has been raised regarding environmental impacts. These concerns centered initially on water quality issues, including the potential contamination of ground water and surface water during hydraulic fracturing activities. They have since incorporated other issues, such as water management practices (both consumption and discharge), land use changes, induced seismicity, infrastructure requirements, and emissions of air pollutants. The array of new questions over hydraulic fracturing in unconventional reservoirs has led, in part, to various grassroots movements, some political opposition, and calls for additional regulatory actions and/or moratoria at the local, state, and federal levels.

As with any new or innovative technology that has the potential to impact human health and the environment, oversight typically begins at the state and local level. Industry and community stakeholders work together with government agencies to assess the technology's impacts and negotiate proper safeguards. Regional and/or federal oversight may be implemented subsequently if the need for a minimum national standard is deemed appropriate. National standards are often adopted from recognized "best practices" among state and local regulations as well as internal industry practices. Currently, the development of natural gas in the United States, including unconventional natural gas, is regulated under a complex set of local, state, and federal laws that addresses many—but not all—aspects of exploration, production, and distribution. State and local authorities are responsible for virtually all of the day-to-day oversight of natural gas development. The organization of this oversight within each gas producing jurisdiction varies considerably. In general, each state has one or more regulatory agencies that may permit wells, including their design, location, spacing, operation, and abandonment, and may regulate for environmental compliance, including water management and disposal, air emissions, underground injection, wildlife impacts, surface disturbance, and worker health and safety.

Federal laws govern some environmental aspects of natural gas development. Where applicable, these controls are largely implemented by state and local authorities. For example, the Clean Water Act (CWA) regulates surface discharges of water associated with natural gas drilling and production, as well as contaminated storm water runoff from production sites. The Safe Drinking Water Act (SDWA) regulates the underground injection of fluids used in hydraulic fracturing if the fluids contain diesel fuel.[6] The National Environmental Policy Act (NEPA) requires that exploration and production on federal lands be assessed for environmental impacts.[7] The Clean Air Act (CAA) limits emissions from associated engines and gas processing equipment. New CAA standards, promulgated on August 16, 2012, expand this coverage to include some natural gas extraction and production processes.[8] The U.S. Environmental Protection Agency (EPA) administers most of the federal pollution control laws, although development on federally owned land is managed primarily by the Bureau of Land Management, in the Department of the Interior, and the U.S. Forest Service, in the Department of Agriculture,[9] and interstate pipelines are regulated by the Federal Energy Regulatory Commission.

THE NATURAL GAS INDUSTRY

Natural gas is a non-renewable, fossil fuel that is used both as an energy source (for heating, transportation, and electricity generation) and as a manufacturing ingredient (for such varied products as plastic, fertilizer, anti-freeze, and fabrics). Natural gas is recovered primarily through drilling and extraction operations by the crude oil and natural gas industry.[10] This industry includes operations in the extraction and production of crude oil and natural gas, as well as the processing, transmission, and distribution of natural gas. For both operational and regulatory reasons, the sector is commonly separated into four major segments: (1) crude oil and natural gas production,[11] (2) natural gas processing, (3) natural gas transmission and storage, and (4) natural gas distribution (see Figure 1). Below is a brief outline of the crude oil and natural gas industry.[12] For more detail regarding specific activities or equipment, see the glossary of terms provided in Table C-1.

Production (Upstream). Production operations include the wells and all related processes used in the extraction, production, recovery, lifting, stabilization, separation, and treating of oil and/or natural gas (including condensate). Production operations span the initial well drilling, hydraulic fracturing, well completion, and workover activities and cover all the portable non-self-propelled apparatus associated with those operations. Production sites include not only the "pads" where the wells are located, but also the stand-alone sites where oil, condensate, produced water, and gas from several wells may be separated, stored, and treated, as well as the low pressure, small diameter, gathering pipelines and related components that collect and transport the oil, gas, and other materials and wastes from the wells to the refineries or natural gas processing plants.

Processing (Midstream). Natural gas is primarily made up of methane. However, in its raw state, natural gas is a mixture of various hydrocarbons and may contain trace amounts of other chemical substances that must be removed before distribution. The additional hydrocarbons are often referred to as natural gas liquids (NGL). They are sold separately and have a variety of different uses. Raw natural gas may also contain water vapor, non-hydrocarbon compounds, and other chemical substances. Processing operations are used to separate out the additional components from raw natural gas to produce "pipeline quality" or "dry" natural gas for consumption.

Transmission and Storage (Downstream). The pipeline quality natural gas leaves the processing segment and enters the transmission segment. Pipelines in the natural gas transmission segment can be interstate pipelines, that carry natural gas across state boundaries, or intrastate pipelines, that transport the gas within a single state. While interstate pipelines may be of a larger diameter and operate at a higher pressure, the basic components are the same. To ensure that the natural gas flowing through any pipeline remains pressurized, compressor stations are required at regular intervals. Further, to ensure proper load balancing during the delivery and receipt of natural gas, the transmission segment often includes underground storage facilities, typically consisting of both man-made and natural sites, such as depleted gas reservoirs and/or salt dome caverns. Pipelines can also be used as storage.

Distribution. The distribution segment is the final step in delivering natural gas to customers. The natural gas enters the distribution segment from delivery points located on interstate and intrastate transmission pipelines and then flows to business and household customers. Nationwide, natural gas distribution systems consist of thousands of miles of piping, including mains and service lines to the customers, as well as compressor and metering stations, which allow distribution companies to both move and monitor the natural gas in the system. The delivery point where the natural gas leaves the transmission segment and enters the distribution segment is often called the "citygate." Typically, the citygate serves as the transfer point of ownership from producers to utilities.

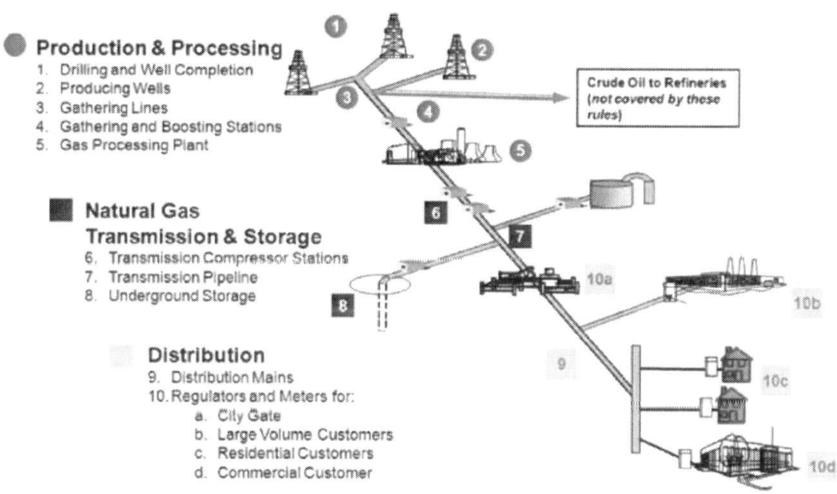

Source: U.S. Environmental Protection Agency.

Figure 1. The Natural Gas Industry.

AIR QUALITY ISSUES IN NATURAL GAS SYSTEMS

Raw natural gas is a mixture of various hydrocarbons (primarily methane) and may contain trace amounts of other chemical substances that must be removed before distribution. Air pollutants associated with the natural gas industry may be emitted through the release of raw natural gas (either purposefully or accidently), the combustion of raw natural gas (either for use or for safety/disposal), the combustion of other fuel resources (for process heat, power, and electricity), and the discharge of particulate matter during construction, transportation, and associated operations. Pollutants include, most prominently, methane and volatile organic compounds, of which the natural gas industry is one of the highest emitting industrial sectors in the United States.[13] Pollutants also include nitrogen oxides, sulfur dioxide, particulate matter, and various forms of hazardous air toxics, including n-hexane, the BTEX compounds (i.e., benzene, toluene, ethylbenzene, and xylene), and hydrogen sulfide. Sources of emissions include pad, road, and pipeline construction; drilling, completion, and flowback activities that occur during the development of a well; and gas processing and transmission equipment such as controllers, compressors, dehydrators, pipelines, and storage vessels.

Emissions

Emissions from the natural gas industry are released in a variety of ways. This report categorizes three types of emissions: *fugitive, combusted*, and *associated.*[14]

Fugitive

"Fugitive" refers to emissions of raw natural gas. Fugitive emissions are commonly defined as either intentional (i.e., vented) or unintentional (i.e., leaked).[15] Intentional emissions are those which are designed specifically into the system (e.g., emissions released to the atmosphere through venting or blow-downs used to guard against over-pressuring; or through gas-driven equipment used to power activities or regulate pressure). Conversely, unintentional emissions result from uncontrolled leaks in the system (e.g., emissions released to the atmosphere through routine wear, tear, and corrosion; improper installation or maintenance; or the overpressure of gases or liquids in the system). Fugitive natural gas is primarily a mixture of low molecular-weight hydrocarbon compounds that are gases at surface pressure and temperature conditions. While the principal component of natural gas is methane (CH4), it may contain smaller amounts of other hydrocarbons, such as ethane, propane and butane, as well as trace amounts of heavier hydrocarbons. These heavier hydrocarbons may include types of volatile organic compounds (VOCs), classified by EPA as ozone precursors and/or as hazardous (i.e., toxic) air pollutants (HAPs). Non-hydrocarbon gases, such as carbon dioxide (CO2), helium (He), hydrogen sulfide (H2S), nitrogen (N2), and water vapor (H2O), may also be present in any proportion to the total hydrocarbon content. The chemical composition and quality of raw natural gas varies across resource reservoirs, and the gas may or may not be "associated" with oil resources. When natural gas is found to be primarily methane, it is referred to as "dry" or "pipeline quality" gas. When natural gas is found bearing higher percentages of heavier hydrocarbons, non-hydrocarbon gases, and/or water vapor, it is commonly referred to as "wet" or "hot" gas. Similarly, quantities of VOCs, HAPs, and hydrogen sulfide can vary significantly depending upon the resource reservoir. VOCs and HAPs compositions typically account for only a small percentage of natural gas mixtures; however, this ratio increases the "wetter" the gas. Natural gas mixtures with a higher percentage of hydrogen sulfide are generally referred to as "sour" or "acid" gas.[16\]

Combusted

"Combusted" refers to emissions of combusted natural gas and other fossil fuels. Combusted emissions are commonly released through (1) the flaring of natural gas for safety and health precautions or (2) the combustion of fossil fuels used to provide process heat, power, and electricity for the system. Flaring is a means to eliminate natural gas that may be impracticable to use, capture, or transport. As with venting, the primary purpose of flaring is to act as a safety device to minimize explosive conditions. Gas may be flared at many points in the system, however, it is most common during the drilling and well completion phases. In addition to flaring, combustion related emissions are released through the use of diesel, gasoline, coal, and natural gas as fuel or power in production, processing, transmission, storage, and distribution equipment (e.g., drills, heaters, engines, furnaces, and construction and transportation machinery). Combusted natural gas releases carbon dioxide (CO_2), carbon monoxide (CO), nitrogen oxides (NOx), and trace amounts of sulfur dioxide (SO_2), particulate matter (PM), uncombusted methane (CH_4), VOCs, and HAPs.

Natural gas combustion is generally considered a greater pollution control mechanism than the venting of natural gas, because the process serves to incinerate many of the VOCs and HAPs that would otherwise be released directly into the atmosphere.

Similarly, natural gas combustion is generally considered as "cleaner" than other fossil fuel combustion with respect to various criteria pollutants and greenhouse gas (GHG) emissions.[17]

Associated

"Associated" refers to secondary sources of emissions that arise from associated operations in crude oil and natural gas production. Associated emissions—including particulate matter—may result from vehicle exhaust, dust, or soil entering the air during construction, transportation, drilling, or well completion.

The Focus of This Report Is on Fugitive Emissions

While there may be significant emissions from the natural gas sector as a result of the combustion of fossil fuels for process heat, power, and transportation, as well as the associated release of particulate matter from construction and road use, the primary focus of this report is on air quality issues related to the resource itself (i.e., the fugitive release of raw natural gas). It is the fugitive release of raw natural gas—and the pollutants contained within it—that makes air quality considerations in the crude oil and natural gas sector unique from other industrial-, construction-, and transportation-intensive sectors.

Pollutants

Through provisions in the Clean Air Act (CAA),[18] the U.S. Environmental Protection Agency (EPA) classifies air pollutants under several different categories, including:

- **Criteria Pollutants.** Common emissions that can harm human health or the environment, or cause property damage, including ground-level ozone (i.e., smog), nitrogen oxides, sulfur dioxide, carbon monoxide, particulate matter, and lead,[19]
- **Hazardous Air Pollutants.** Toxic chemicals that are known or suspected to cause cancer or other serious health effects, such as reproductive diseases or birth defects,[20]
- **Greenhouse Gas Pollutants.** Chemical compounds that trap heat in the atmosphere and contribute to the forcing of climate change.[21]

The Criteria and Hazardous Air Pollutant categories are identified and regulated under specific provisions of the CAA. In contrast, greenhouse gases may be regulated under several different ones.[22] The natural gas industry produces emissions from each of these categories, some of which may overlap (e.g., benzene is a volatile organic compound [VOCs]; and, it is considered both a ground-level ozone-producing criteria pollutant and a carcinogenic hazardous air pollutant [HAPs]). See Figure 2 for a comparison of air pollution emissions from the natural gas industry against those of other U.S. industrial sectors.

Volatile Organic Compounds (VOCs)—A Ground-level Ozone (O_3)[23] Precursor

The natural gas industry is currently one of the largest sources of VOCs emissions in the United States, accounting for over 67% of the VOCs emissions from all industrial source categories (and, approximately 12% from all domestic source categories).[24] VOCs—in the form of various hydrocarbons—are emitted throughout a wide range of crude oil and natural gas production operations and equipment. The interaction between VOCs and NO_x in the atmosphere contributes to the formation of ozone (i.e., smog). EPA quantifies several health effects associated with exposure to ozone, including premature death, heart failure, chronic respiratory damage, and premature aging of the lungs. Ozone may also exacerbate existing respiratory illnesses, such as asthma and emphysema, or cause chest pain, coughing, throat irritation, and congestion.[25]

Nitrogen Oxides (NO_x)—A Ground-Level Ozone (O_3) Precursor

Significant amounts of NO_x are emitted at natural gas sites through the combustion of fossil fuels. This combustion occurs during several activities, including (1) the flaring of natural gas during drilling and well completions, (2) the combustion of fuels in drills, heaters, boilers, and other production, construction, and transportation equipment, and (3) the combustion of fuels in engines and turbines to drive the compressors that move natural gas through the system.[26]

In addition to ozone formation (see VOCs description above), current scientific evidence links short-term NO_x exposures with adverse respiratory effects including airway inflammation in healthy people and increased respiratory symptoms in people with asthma.

Carbon Monoxide (CO)

Similar to NO_x, CO is emitted from combustion processes in stationary and mobile sources. CO can cause harmful health effects by reducing oxygen delivery to the body's organs (like the heart and brain) and tissues.

Particulate Matter (PM)

PM may occur from dust or soil entering the air during well-pad construction, traffic on access roads, and diesel exhaust from drilling machinery, vehicles, and other engines. PM is linked to respiratory and cardiovascular problems, including aggravated asthma attacks, chronic bronchitis, decreased lung function, heart attacks, and premature death.[27]

Sulfur Dioxide (SO_2)

SO_2 is emitted from crude oil and natural gas production and processing operations that handle and treat sulfur-rich, or "sour," gas. Current scientific evidence links short-term exposures to SO_2 with an array of adverse respiratory effects including bronchoconstriction and increased asthma symptoms.

Hazardous Air Pollutants (HAPs)

HAPs, also known as air toxics, are those pollutants that are known or suspected to cause cancer or other serious health effects, such as reproductive diseases, or birth defects. Of the HAPs emitted from natural gas production operations, the chemicals referred to as volatile organic compounds (VOCs) are the largest group, and typically evaporate easily into the air. The most common HAPs in natural gas systems are n-hexane, the BTEX compounds (benzene, toluene, ethylbenzene, and xylenes), and hydrogen sulfide.[28] HAPs are found primarily in natural gas itself, and are emitted from equipment leaks and from various processing, compressing, transmission, distribution, or storage operations. They are also a byproduct of fuel combustion and may be components in various chemical additives.

Methane (CH_4)

Methane is the principal component of natural gas. Methane is also a precursor to ground-level ozone formation[29] and a potent greenhouse gas (GHG), albeit with a shorter climate-affecting time horizon than carbon dioxide.[30] Every process in the natural gas production, processing, transmission and storage, and distribution sectors has the potential to emit significant amounts of methane.

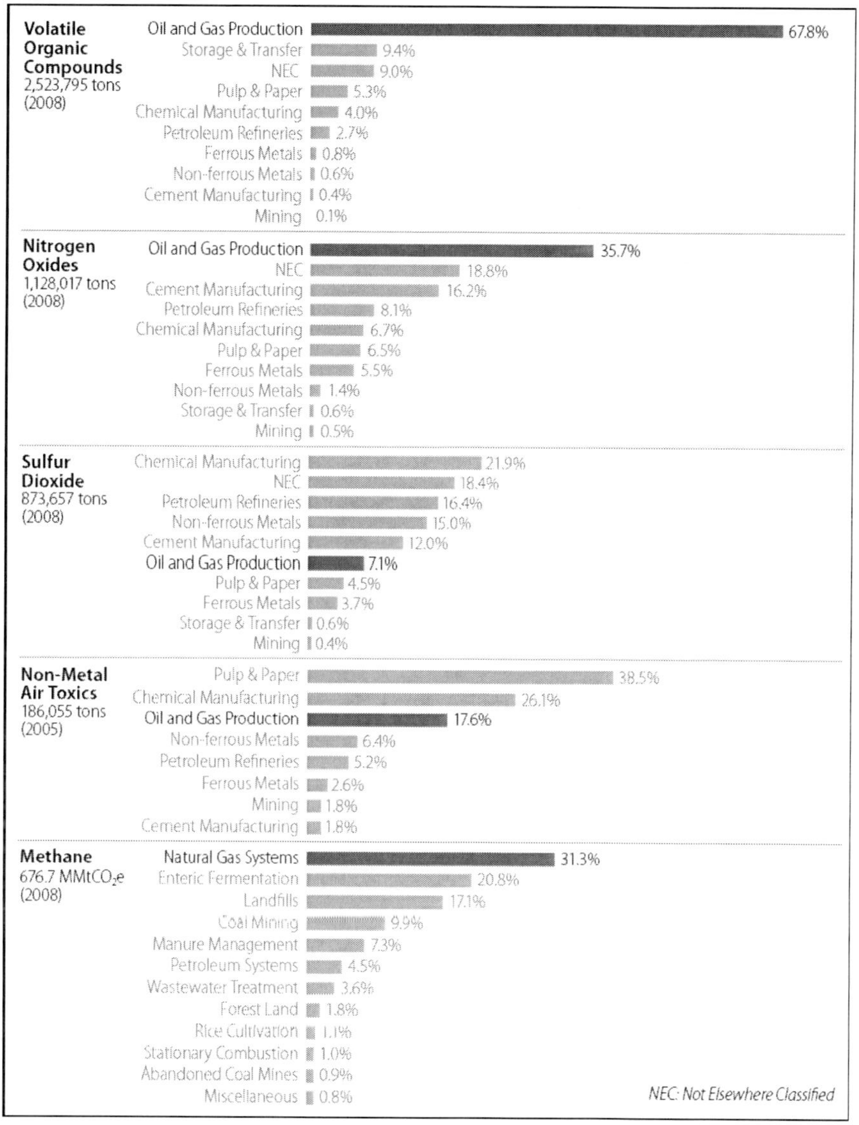

Source: EPA, National Emissions Inventory 2008, v.1.5 GPR, http://www.epa.gov/ttnchie1/net/ 2008inventory.html; National Air Toxics Assessment, 2005, http://www.epa.gov/ttn/atw/natamain/; Inventory of U.S. Greenhouse Gas Emissions and Sinks: 1990-2009, http://www.epa.gov /climatechange/ghgemissions/ usinventoryreport/archive.html.

Figure 2. Selected Emissions Inventories for U.S. Industrial Sectors.

EPA's *Inventory of U.S. Greenhouse Gas Emissions and Sinks: 1990-2010* (published April 15, 2012) estimates 2010 methane emissions from "Natural Gas Systems" to be 215.4 million metric tons of carbon dioxide equivalent (MMtCO2e).[31] In each year since 1990, natural gas systems were cited as being the single largest contributor to U.S. anthropogenic (i.e., man-made) methane emissions. In 2010, natural gas systems represented nearly a third of the total methane emissions from all domestic sources and accounted for about 3% of all GHG emissions in the United States.[32] As a GHG, methane contributes to climate change, which over time may lead to increased air and ocean temperatures, changes in precipitation patterns, melting and thawing of glaciers and ice sheets, increasing intensity and severity of weather events, and sea level rise, among other impacts.[33]

Sources

Natural gas systems include many activities and pieces of equipment that have the potential to emit air pollutants. Most of these emissions sources are common to both conventional and unconventional natural gas development, with the most significant exception being high-volume, slick-water hydraulic fracturing activities and their flowback.

Drilling

Fugitive natural gas and other air pollutants may escape to the atmosphere during initial drilling operations through the circulation of drilling fluids back to the surface. Further, emissions from combusted natural gas—including CO_2, NO_x, and potentially SO_2—may also be released at the well site due to both engine combustion and flaring activities. These include "well test flaring," which occurs during the drilling and testing of oil and gas wells, and "solution gas flaring," which occurs during the disposal of associated gas produced along with crude oil (as is the case currently with oil production in the Bakken formation).

Well Completions

Well completions contain several processes which have the potential to emit air pollutants. For example, hydraulic fracturing uses pressurized fluids containing any combination of water, proppant, and added chemicals to penetrate and produce natural gas from tight formations (e.g., shale, sand, or coal formations). The process requires a high rate, extended flowback period to expel fracture fluids and solids from the well.[34] Reservoir gas may escape with these fluids during flowback or impoundment, allowing emissions of methane and VOCs to be released to the atmosphere. Conversely, if the flowback gas is separated from the fluids and flared, emissions of nitrous oxides and carbon dioxide may be released to the atmosphere. EPA estimates that well completions involving hydraulic fracturing can vent substantially more natural gas—approximately 230 times more—than well completions not involving hydraulic fracturing,[35] if not controlled by reduction equipment.[36] Other sources from industry and academia estimate these emissions levels differently, in part because the data are limited.[37]

Compressors

There are many locations throughout the natural gas sector where compression is required to move gas along the pipeline. This is accomplished by different machinery such as combustion turbines, reciprocating internal combustion engines, and electric motors. Both the turbine-powered centrifugal compressors and the reciprocating internal combustion compressors may use a small portion of the natural gas they compress to fuel the turbine. Both are potential sources of VOCs and methane emissions. Centrifugal compressors require seals around the rotating shaft to prevent gases from escaping where the shaft exits the casing. The seals in some compressors use oil (e.g., "wet seal compressors"), and they commonly vent the absorbed gas to the atmosphere when they are purged. Reciprocating compressors, on the other hand, leak natural gas throughout the course of their normal operation, with the highest volume of gas loss associated with worn down piston rod packing systems. Using dry-seal centrifugal systems or periodically replacing worn down rod packing systems are the most effective ways of controlling emissions from gas-driven compressors. Conversely, where available, electric motors can be used to operate compressors. This type of compression does not require the use of any of the natural gas from the pipeline, but it does require a source of electricity.

Controllers

Pneumatic controllers are automated instruments widely used in the natural gas sector for maintaining pressure, temperature, and flow rate conditions in the system. In many situations, the pneumatic controllers make use of the available high-pressure natural gas in the system to regulate these conditions. In these "gas-driven" pneumatic controllers, natural gas may be released intermittently with every valve movement or continuously from the valve control pilot. Gas driven pneumatic controllers are typically characterized as either “high-bleed” or “low-bleed,” where a high-bleed device releases at least 6 cubic feet of gas per hour. Conversely, “non-gas driven” pneumatic controllers use sources of power other than pressurized natural gas, greatly reducing levels of emissions. Examples include solar, electric, and instrument air.

Storage Vessels

After being separated from the natural gas stream, crude oil, condensate, and produced water are typically stored in fixed-roof storage vessels. These vessels, which are operated at or near atmospheric pressure conditions, can release various emissions to the atmosphere as a result of working, breathing, and flash losses. Working losses occur due to the emptying and filling of storage tanks. Breathing losses are the release of gas associated with daily temperature fluctuations and other equilibrium effects. Flash losses occur when a liquid is transferred from a vessel with higher pressure to a vessel with lower pressure, thus allowing entrained gases or a portion of the liquid to vaporize or flash. Typically, the larger the pressure drop, the more flash emissions will occur in the storage stage. The two ways of controlling tanks with significant emissions are to install vapor recovery units (VRU) or to route the emissions from the tanks to control devices (i.e., flares).

Dehydrators

Once natural gas has been separated from any liquid materials or products (e.g., crude oil, condensate, or produced water), residual entrained water is removed from the natural gas by dehydration. One of the most widely used natural gas dehydration processes is glycol dehydration. Glycol dehydration is an absorption process in which a liquid absorbent (glycol) directly contacts the natural gas stream and absorbs both the entrained water vapor as well as a number of selected hydrocarbons, including BTEX, n-hexane, and other HAPs emissions. During the recirculation stages, the hydrocarbons are boiled off along with the water and are either vented to the atmosphere or directed to a control device.

General Equipment and Pipeline Leaks

Fugitive emissions can emanate from valves, pump seals, flanges, compressor seals, pressure relief valves, open-ended lines, and other process and operation components at any point during operations. Leaks may be due to routine wear, tear, and corrosion; improper installation or maintenance; or the overpressure of gases or liquids in the system. Because of the large number of valves, pumps, and other components within a natural gas production, processing, or transmission facility, equipment leaks collectively can be a significant source of emissions. Further, there are over 300,000 miles of transmission pipelines alone in the United States, and these pieces of equipment exist throughout the system.

Workovers, Maintenance, and Upsets

Periodically, wells require re-stimulation workovers or routine maintenance—such as the unloading of liquids or blowdowns—in order to reestablish productive gas flows. These activities, depending upon their methods, may release significant amounts of fugitives into the atmosphere. Further, all activities are susceptible to occasional upsets and accidental losses.

THE CLEAN AIR ACT AND THE FEDERAL ROLE IN AIR QUALITY ISSUES IN NATURAL GAS SYSTEMS

The Clean Air Act (CAA)[38] seeks, in part, to protect human health and the environment from emissions that pollute ambient, or outdoor, air.[39] It requires the U.S. Environmental Protection Agency (EPA) to establish minimum national standards for air emissions from various source categories (sources in the "Crude Oil and Natural Gas Production," and the "Natural Gas Transmission and Storage" sectors are included under these categories), and assigns primary responsibility to the states to assure compliance with the standards. EPA has largely delegated day-to-day responsibility for CAA implementation to all 50 states, including permitting, monitoring, inspections, and enforcement; and in many cases, states have further delegated program implementation to local governments. Sections of the CAA which are most relevant to air quality issues in the natural gas industry are discussed below. Some of the regulations implementing these sections have been revised by new federal air standards promulgated by EPA on August 16, 2012. The paragraphs below summarize relevant sections of the CAA prior to these revisions. For a summary of the new air standards and their relationship to the prior ones, see the subsequent section, "New Federal Air Standards for Crude Oil and Natural Gas Systems," as well as Appendix A.

Prior Federal Air Standards and Other Rules for Crude Oil and Natural Gas Systems

National Ambient Air Quality Standards (NAAQS)

Section 109 of the CAA requires EPA to establish NAAQS for air pollutants that may reasonably be anticipated to endanger public health or welfare, and whose presence in ambient air results from numerous or diverse sources. Using this authority, EPA has promulgated NAAQS for sulfur dioxide (SO_2), particulate matter ($PM_{2.5}$ and PM_{10}), nitrogen dioxide (NO_2), carbon monoxide (CO), ozone, and lead. States are required to implement specified air pollution control plans to monitor these pollutants and ensure the NAAQS are met, or "attained." Additional measures are required in areas not meeting the standards, referred to as "nonattainment areas." "Nonattainment" findings for ground-level ozone, nitrogen oxides, and sulfur dioxide in areas with crude oil and natural gas production may result in states establishing specific pollution controls that could affect the industry.

Permits

The CAA Amendments of 1990 added Title V (i.e., "5")[40] which requires states to administer a comprehensive permit program for the operation of sources emitting air pollutants. Sources subject to the permit requirements generally include new or modified "major sources" that emit or have the potential to emit 100 tons per year of any regulated pollutant, plus new or existing "area sources" that emit or have potential to emit lesser specified amounts of hazardous air pollutants. In "nonattainment" areas, the permit requirements may include sources which emit as little as 50, 25, or 10 tons per year of VOCs, depending on the severity of the region's nonattainment status ("serious," "severe," or "extreme" respectively). A Title V permit must, among other things, list all emissions limitations and standards applicable to the source, ensure that monitoring and recordkeeping are sufficient to demonstrate compliance, and require the payment of fees. Currently, most crude oil and natural gas production activities upstream from the processing plant are not classified as "major sources" under EPA Title V operating permits.[41]

Greenhouse Gas Reporting

In the FY2008 Consolidated Appropriations Act (H.R. 2764; P.L. 110-161), Congress directed EPA to develop regulations that establish a mandatory GHG reporting program that applies "above appropriate thresholds in all sectors of the economy." EPA issued the Mandatory Reporting of Greenhouse Gases Rule (MRR)[42] which became effective on December 29, 2009, and included reporting requirements for facilities and suppliers in 32 source categories.[43] Affected facilities in the petroleum and natural gas industry include onshore petroleum and natural gas production, offshore petroleum and natural gas production, natural gas processing, natural gas transmission compressor stations, underground natural gas storage, liquefied natural gas (LNG) storage, LNG import and export terminals, and natural gas distribution. The rule requires petroleum and natural gas facilities that emit 25,000 metric tons or more of carbon dioxide (CO_2) equivalent per year to report annual fugitive emissions of methane and CO_2 from equipment leaks and venting, and annual combusted emissions of CO_2, methane, and nitrous oxide from gas flaring, from stationary and portable equipment involved in onshore petroleum and natural gas production, and from stationary equipment involved in natural gas distribution. Since methane is the most prominent emission from upstream oil and gas activities, and since the rule requires producers to aggregate emissions from all commonly controlled wells and their associated equipment (e.g., compressors, generators, piping, and storage tanks), many facilities may likely be required to report GHG emissions in coming years.[44] EPA estimates that the rule covers 85% of the total GHG emissions from most of the U.S. petroleum and natural gas industry with approximately 2,800 facilities reporting. Many states have their own GHG reporting requirements independent of the 2009 federal rules (e.g., the Regional GHG Initiative, the Western Climate Initiative, and California).

New Source Performance Standards (NSPS)

Section 111 of the CAA requires EPA to promulgate regulations establishing emissions standards that are applicable to new, modified, and reconstructed sources—if such sources cause or contribute significantly to air pollution that may reasonably be anticipated to endanger public health or welfare. A performance standard reflects the degree of emission limitation achievable through the application of the "best system of emission reduction" (BSER) which EPA determines has been adequately demonstrated. As new technology advances are made, EPA is required to revise and update NSPS applicable to designated sources.

The following federal NSPS may apply to crude oil and natural gas systems (some parts have been rewritten by the April 17, 2012, standards, see the next section and **Appendix A** for further discussion):

- 40 C.F.R. Part 60, Subpart JJJJ—Standards of Performance for Stationary Spark Ignition (SI) Internal Combustion Engines (ICE). Subpart JJJJ applies to manufacturers, owners, and operators of SI ICE, which affects new, modified, and reconstructed stationary SI ICE (i.e., generators, pumps, and compressors), combusting any fuel (i.e., gasoline, natural gas, LPG, landfill gas, digester gas etc.), except combustion turbines. The applicable emissions standards are based on engine type, fuel type, and manufacturing date. The regulated pollutants are NO_x, CO, and VOCs, and there is a sulfur limit on gasoline. Subpart JJJJ would apply to facilities operating spark ignition engines at compressor stations;
- 40 C.F.R. Part 60, Subpart IIII—Standards of Performance for Stationary Compression Ignition (CI) ICEs. Subpart IIII applies to manufacturers, owners, and operators of CI ICE (diesel), which affects new, modified, and reconstructed (commencing after July 11, 2005) stationary CI ICE (i.e., generators, pumps, and compressors), except combustion turbines. The applicable emissions standards (phased in Tiers with increasing levels of stringency) are based on engine type and model year. The regulated pollutants are NO_x, PM, CO, and non-methane hydrocarbons (NMHC), while the emissions of sulfur oxides (SO_x) are reduced through the use of low sulfur fuel. Particulate emissions are also reduced by standards. Subpart IIII would apply to facilities operating compression ignition engines at compressor stations;
- 40 C.F.R. Part 60, Subpart KKK—Standards of Performance for Equipment Leaks of VOCs from Onshore Natural Gas Processing Plants. Subpart KKK applies to gas processing plants that are engaged in the extraction of natural gas liquids from field gas and contains provisions for VOCs leak detection and repair (LDAR) (this section was revised by the new air standards);
- 40 C.F.R. Part 60, Subpart LLL—Standards of Performance for Onshore Natural Gas Processing: SO_2 Emissions. Subpart LLL governs emissions of SO_2 from gas processing plants, specifically gas sweetening units (which remove H_2S and CO_2 from sour gas) and sulfur recovery units (which recover elemental sulfur) (this section was revised by the new air standards); and
- 40 C.F.R. Part 60 Subpart Kb—Standards of Performance for Volatile Organic Liquid Storage Vessels (Including Petroleum Liquid Storage

Vessels) for which construction, reconstruction, or modification commenced after July 23, 1984.

National Emission Standards for Hazardous Air Pollutants (NESHAPs)

Section 112 of the CAA requires EPA to promulgate regulations establishing standards to control emissions of hazardous air pollutants (HAPs). NESHAPs are applicable to both new and existing sources of HAPs, and there are NESHAPs for both "major" sources of HAPs and "area" sources of HAPs. A major source of HAPs is one with the potential to emit in excess of 10 tons per year (tpy) of any single HAPs or 25 tpy of all HAPs, combined. An area source of HAPs is a stationary source of HAPs that is not major. The aim is to develop technology-based standards which require levels met by the best existing facilities (commonly referred to as maximum achievable control technology [MACT] standards). The pollutants of concern in the oil and gas sector are primarily the BTEX compounds, formaldehyde, and n-hexane.

The following federal NESHAPs may apply to crude oil and natural gas systems (some parts have been rewritten by the April 17, 2012, standards, see the next section and Appendix A for further discussion):

- 40 C.F.R. Part 63, Subpart ZZZZ—National Emission Standards for Hazardous Air Pollutants for Reciprocating Internal Combustion Engines (RICE);
- 40 C.F.R. Part 63, Subpart H—National Emission Standards for Organic Hazardous Air Pollutants for Equipment Leaks. Subpart H applies to equipment that contacts fluids with a HAPs concentration of 5%;
- 40 C.F.R. Part 63, Subpart HH—NESHAPs from Oil and Natural Gas Production Facilities. Subpart HH controls air toxics from oil and natural gas production operations and contains provisions for both major sources and area sources of HAPs. Emission sources affected by this regulation are tanks with flash emissions (major sources only), equipment leaks (major sources only), and glycol dehydrators (major and area sources) (this section was revised by the new air standards);
- 40 C.F.R. Part 63, Subpart HHH—NESHAPs from Natural Gas Transmission and Storage Facilities. Subpart HHH controls air toxics from natural gas transmission and storage operations. It affects glycol dehydrators located at major sources of HAPs (this section was revised by the new air standards); and

- 40 C.F.R. Part 61, Subpart V—National Emission Standard for Equipment Leaks (Fugitive Emission Sources). Subpart V applies to equipment that contacts fluids with a volatile HAPs concentration of 10%.

New Federal Air Standards for Crude Oil and Natural Gas Systems (August 16, 2012)

On January 14, 2009, two non-governmental organizations filed a complaint under the citizen suit provision of the CAA, alleging that EPA failed to meet its obligations under CAA Sections 111(b)(1)(B), 112(d)(6) and 112(f)(2) to take actions relative to the review/revision of the NSPS and the NESHAPs with respect to the "Crude Oil and Natural Gas Production" source category.[45] On February 4, 2010, the U.S. Court of Appeals for the D.C. Circuit entered a consent decree requiring EPA to sign proposed standards and/or determinations not to issue standards by January 31, 2011 (modified to July 28, 2011), and to take final action by November 30, 2011 (modified to April 17, 2012). EPA proposed a new set of air standards for the "Crude Oil and Natural Gas Production" sector and the "Natural Gas Transmission and Storage" sector on July 28, 2011,[46] and held three public hearings for the proposal. After several court-agreed extensions, the final rules establishing the new standards were signed by the Administrator on April 17, 2012, were published in the *Federal Register* on August 16, 2012,[47] and became effective on October 15, 2012. A summary of the new federal air standards is provided below. For a detailed comparison of the new federal air standards against the prior federal air standards, see Appendix A.

New NSPS for Crude Oil and Natural Gas Systems

The new NSPS for the "Crude Oil and Natural Gas Production" and the "Natural Gas Transmission and Storage" source categories regulate volatile organic compounds (VOCs) emissions from gas wells, centrifugal compressors, reciprocating compressors, pneumatic controllers, storage vessels, and leaking components at onshore natural gas processing plants, as well as sulfur dioxide (SO_2) emissions from onshore natural gas processing plants. Prior to the new standards, processing plants were the only source category regulated at the federal level. The new standards include:

- **Gas Wells.** The rule covers any gas well that is "an onshore well drilled principally for production of natural gas" and is "hydraulically fractured." Oil wells (i.e., wells drilled principally for the production of crude oil) or conventional gas wells (i.e., wells drilled without hydraulic fracturing for the production of natural gas) are not subject to the rule. For fractured and refractured gas wells, the rule requires owners/operators to use "reduced emissions completions"—also known as "REC" or "green completions"—to reduce VOCs emissions from well completions. A REC is defined by EPA as "a well completion following fracturing or refracturing where gas flowback that is otherwise vented is captured, cleaned, and routed to the flow line or collection system, re-injected into the well or another well, used as an on-site fuel source, or used for other useful purpose that a purchased fuel or raw material would serve, with no direct release to the atmosphere" (see graphic in **Figure 3**). To provide industry enough time to order and manufacture the necessary REC equipment, the NSPS establishes two phases for compliance. Owners and/or operators may use either REC or completion combustion devices (e.g., flaring) until January 1, 2015. After January 1, 2015, REC will be required. The rule exempts exploratory, delineation, and low-pressure gas wells from the REC requirement, stipulating the use of completion combustion devices instead.[48]
- **Storage Vessels.** The rule requires individual storage vessels in the crude oil and natural gas production segment and the natural gas processing, transmission, and storage segments with emissions equal to or greater than 6 tons per year (tpy) to achieve at least 95% reduction of uncontrolled VOCs emissions.
- **Certain Controllers.** The rule sets a natural gas bleed rate limit of 6 standard cubic feet per hour (scfh) for individual, continuous bleed, natural gas-driven pneumatic controllers located between the wellhead and the point at which the gas enters the transmission and storage segment. For individual, continuous bleed, natural gas-driven pneumatic controllers located at natural gas processing plants, the rule sets a natural gas bleed limit of zero scfh.
- **Certain Compressors.** The rule requires a 95% reduction of VOCs emissions from wet seal centrifugal compressors located between the wellhead and the point at which the gas enters the transmission and storage segment. The rule also requires measures intended to reduce VOCs emissions from reciprocating compressors located between the

wellhead and the point where natural gas enters the natural gas transmission and storage segment. Owners and/or operators of these compressors must replace the rod packing of these reciprocating compressors based on specified usage or time.

- **Onshore Natural Gas Processing Plants.** The rule revises the existing NSPS requirements for leak detection and repair (LDAR) to reflect the procedures and leak thresholds established in the NSPS for equipment leaks of VOCs emissions in the synthetic organic chemicals manufacturing industry. This rule also revises the existing NSPS requirements for SO_2 emission reductions based on sulfur feed rate and sulfur content of gas.

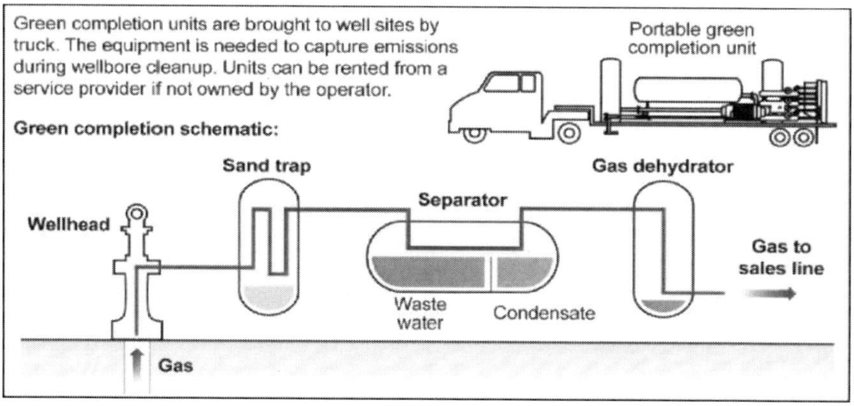

Source: U.S. Environmental Protection Agency.
Graphic: Dave Merrill/BGOVgraphics, Bloomberg Government, http://www.bgov.com.

Figure 3. Reduced Emissions Completion Equipment.

New NESHAPs for Crude Oil and Natural Gas Systems

The new rules revise the NESHAPs for glycol dehydration unit process vents and leak detection and repair (LDAR) requirements and retain the existing NESHAPs for storage vessels. The new standards include:

- **Glycol Dehydration Units.** The rule establishes MACT standards for "small" glycol dehydration units, which were unregulated under the initial NESHAPs. Covered glycol dehydrators now include those with an actual annual average natural gas flow rate less than 85,000 standard cubic meters per day (scmd) or actual average benzene emissions less than 0.9 megagrams per year (Mg/yr), and they must meet unit-specific limits for benzene, toluene, ethylbenzene, and xylene (collectively, "BTEX").[49]

- **Leak Detection and Repair.** The rule lowers the leak definition for valves at natural gas processing plants from 10,000 parts per million (ppm) to 500 ppm for major sources at crude oil and natural gas production facilities, thus requiring the application of LDAR procedures at this level.

Effects of the New Federal Air Standards on State Attainment Planning and Permitting

With the release of the new air standards for the crude oil and natural gas production, transmission, and storage sectors, EPA constituted a federally required minimum level of control for various source categories. States have the flexibility to put their own programs in place or implement existing programs as long as they are at least as protective as the federal standards.

Attainment Planning

EPA recently designated attainment and nonattainment areas for the 2008 ozone National Ambient Air Quality Standards (NAAQS). Some of these areas have significant crude oil and natural gas activities. States with ozone nonattainment areas are required to submit modified state implementation plans (SIP) in 2015 and to attain the standard by 2015 and 2018 for areas classified as "marginal" and "moderate," respectively. A few areas classified as "serious" nonattainment must attain by 2021. As the new air standards may likely help states make progress in attaining the ozone NAAQS in nonattainment areas where there is significant well development, states are allowed to include the federal NSPS as a federally enforceable strategy in their nonattainment SIP. States may "take credit" for the NSPS in their SIP towards meeting two requirements: (1) the new standards are expected to achieve 95% control of VOCs emissions from new gas wells, making it easier for states to obtain the overall reduction in emissions they need to attain the ozone NAAQS without adding any federal or state permitting requirements; (2) SIPs in "moderate" and "serious" areas must also show "reasonable further progress" in controlling emissions in the years before they attain the ozone NAAQS. In most areas, states may choose to measure this progress relative to emissions in 2011. In areas that had wells drilled in 2011 and will continue to have more wells drilled in the years ahead, the 95% control from the NSPS may provide emission reductions that can be credited toward the reasonable further progress requirement. In areas that had no or few wells drilled in 2011 but that will see drilling activity in the future, the 95% control from the NSPS may ensure that emissions from new well development do not impede meeting the reasonable further progress requirement.

Permitting

While the NSPS regulates new wells with uncontrolled emissions that are below the existing thresholds that define a "major source" for pre-construction permit and Title V operating permit purposes, EPA has not changed the actual emission thresholds that trigger the requirements for major source permitting. In the absence of the NSPS, some hydraulically fractured gas wells could have emissions above the thresholds that trigger major source pre-construction and Title V permitting requirements in some ozone nonattainment areas. Federally enforceable emissions limits, such as those in the NSPS, are counted when determining whether a well's emissions exceed these applicability thresholds. Wells complying with the NSPS most likely will not trigger major source permitting thresholds. Wells complying with the NSPS may also have emissions low enough to avoid needing a minor source permit from the state, as the final NSPS provides a path (i.e., REC) for existing wells that are refractured to avoid falling under the scope of the NSPS at all, thereby avoiding any automatic requirement to get a state minor source permit. However, states may still include modified wells in their minor source permitting rules if they choose.

Reported Costs and Benefits of the New Federal Air Standards

Natural gas is a product of—and thus a source of revenue for—the oil and gas industry. It is also a main source of pollution for the industry when it is emitted into the atmosphere. Due to this unique linkage, pollution abatement has the potential to translate into economic benefits for the industry, as producers can offset compliance costs with the value of natural gas and condensate recovered and sold at market. EPA reports the environmental and the economic benefits of the new air standards as follows:[50]

- **VOCs Reductions of 190,000 to 290,000 Tons Annually.** VOCs emissions reductions of nearly 95% from hydraulically fractured gas wells are expected to help reduce ground-level ozone (i.e., smog) in areas where crude oil and natural gas production occurs.

- **Air Toxics Reductions of 12,000 to 20,000 Tons Annually.** The new rules are intended to protect against potential cancer risks from emissions of several air toxics, including the BTEX compounds.
- **Methane Reductions of 1.0 Million to 1.7 Million Short Tons Annually.** (Equivalent to 19 to 33 million metric tons of CO_2 equivalent (CO_2e).) While not targeted by the new rules, methane reductions from new and modified well completions and other activities would yield an additional environmental co-benefit.
- **Industry Costs of $170 Million Annually.** EPA estimates the rule will cost producers about $170 million annually in 2015 (in 2008$). Industry and third party sources have estimated anywhere from $450 million to over $2.8 billion, depending upon assumptions regarding the number of wells subject to compliance, the cost of REC equipment and rentals, and the number and cost of completion combustion requirements. (For reference, U.S. natural gas producers sold about 24 trillion cubic feet of gas in 2011, with a value of $72.5 billion at $3/Mcf.)[51]
- **Net Cost Savings for Industry of $11 Million to $19 Million.** EPA estimates that compliance costs would be offset by the sales of the captured methane and natural gas liquids, resulting in a net gain of $11 million to $19 million in 2015. EPA's cost-benefit analysis, as put forth in its *Regulatory Impact Analysis* for the proposed air standards, has been critiqued as both too high and too low by industry and environmental stakeholders, respectively.[52] Industry stresses that the economic analyses must include the full variety of conditions (e.g., the full range of VOCs content found across different reservoirs) in upstream production activities to support all the costs of compliance with the proposed rule; and, that the analysis does not fully take into account the effect that operational standards (i.e., the requirements for additional green completions and other abatement equipment) will have on the production and growth of the industry in the near term.[53] Industry claims that fluctuations in the market price of recovered products may also serve to further depress EPA's revenue estimates. Conversely, environmental stakeholders argue that the costs of control used by EPA are conservative estimates and do not properly account for the full social and environmental benefits created by the capture of methane.[54]

Issues for Congressional Consideration

The oil and gas industry in the United States has grown markedly in recent years. This growth is primarily a result of new or improved technologies (e.g., hydraulic fracturing, directional drilling) used on unconventional resources (e.g., shale gas, coal-bed methane). The expansion, however, has prompted renewed questions about the industry's impacts on human health and the environment, including impacts on air quality. The U.S. Environmental Protection Agency (EPA), in response to a consent decree issued by the U.S. Court of Appeals for the D.C. Circuit, promulgated new air standards for crude oil and natural gas systems on August 16, 2012.[55] These standards—effective October 15, 2012—revised existing rules and promulgated new ones to regulate emissions of volatile organic compounds, sulfur dioxide, and hazardous air pollutants from many source categories that had never before been covered by federal oversight. This expansion in government regulation—in response to the expansion in industry production—has sparked discussion on a number of fronts, with some of the more significant debates involving emissions estimates, impacts, compliance costs, and the role of local, state, and federal governments. These and other outstanding issues for congressional consideration are summarized below.

The Regulatory Role of Federal, State, and Local Governments

- According to EPA, the new air standards are designed to provide minimum federal requirements for emissions of air pollutants in the crude oil and natural gas sector that can both protect human health and the environment and allow for continued growth in production. However, some argue that state and local governments are better positioned to oversee the industry's development. They claim the new standards overlay federal requirements on activities that are already regulated at the state and local level and fail to effectively address the regional and state-specific character of many oil and gas activities, including differences in geology, hydrology, climate, topography, industry characteristics, development history, state legal structures, population density, and local economics. Others, attesting to the "patchwork" of state and local requirements, have called for the federal government to step in and institute consistent nationwide standards for emissions. They claim a federal standard would extend regulatory certainties to the industry and would best ensure health and environmental protections for consumers. In light of these considerations, Congress may choose to re-examine proposed and/or existing federal requirements apropos of existing state and local regulations, or introduce new federal requirements if deemed necessary.

Currently, states lead the day-to-day permitting, monitoring, and enforcement of crude oil and natural gas development. Any federal requirements that might apply have typically been delegated to the states. In general, each state has one or more regulatory agencies that may permit wells, including their design, location, spacing, operation, and abandonment, and may regulate for environmental compliance, including water management and disposal, air emissions, underground injection, wildlife impacts, surface disturbance, and worker health and safety. The organization of regulatory agencies within the various oil and gas producing states varies considerably. In many cases, state agencies were established initially to provide a structure to facilitate—not regulate—oil and gas development. These facilitatory agencies served primarily to provide a central state system for administering resource claims and mediating disputes, not monitoring and overseeing environmental protection. In some instances, these agencies have taken on oversight activities for environmental protection. In other instances, they have operated alongside a separate agency that has been mandated to prevent environmental pollution and public health impacts.

For example, Colorado provides multiple state agencies with different authorities to regulate industry operations. The Colorado Department of Public Health and Environment (CDPHE) and the Colorado Department of Natural Resources maintain separate but complementary oversight of industry operations. Colorado's Air Quality Control Commission (under CDPHE) has regulations to address emissions from tanks, engines, compressors, and associated equipment. Colorado's Oil and Gas Conservation Commission has regulations pertaining to such issues as well completions, odors, noise, and drill rig setbacks. Conversely, Pennsylvania's Department of Environmental Protection (DEP) Office of Oil and Gas Management is responsible for all statewide oil and gas programs. The office manages activities both to facilitate the exploration, development, and recovery of Pennsylvania's oil and gas reservoirs while also overseeing the protection of the commonwealth's natural resources and environment. The office develops policy and programs for the regulation of oil and gas development pursuant to the commonwealth's Oil and Gas Act, the Coal and Gas Resource Coordination Act, and the Oil and Gas Conservation Law. It oversees the oil and gas permitting and inspection programs, develops statewide regulation and standards, conducts training programs for industry, and works with the Interstate Oil and Gas Compact Commission and the Technical Advisory Board.

All crude oil and natural gas producing states have laws in place related to oil and gas development. Most state requirements are written into rules or regulations. Requirements may also be added to permits on a case-by-case basis as a result of findings from environmental reviews, on-the-ground inspections, public comments, or commission hearings. For example, emissions from oil and gas development in Colorado are governed primarily by statutory provisions of the Oil and Gas Conservation Act (Colo. Rev. Stat. §34-60-100, et seq.), Colorado's Air Pollution and Prevention Control Act (§25-7-100, et seq.), and Water Quality Control Act (§25-8-100, et seq.). In other states, such as Montana or Texas, emissions from oil and gas development are addressed most prominently during permitting, registration, and authorization activities (e.g., Montana Department of Environmental Quality, Air Quality Permits [MAQP], under Rule 17.8.752, and Montana Registration Requirements, under Rule 17.8.1711[1][a]; or Texas Commission on Environmental Quality, Permit by Rule [PBR], Standard Permit, and New Source Review [NSR] Permit, under Title 30, Texas Administration Code, Chapter 116, et seq.).

Most state oil and gas regulations were written well before unconventional natural gas development became widespread. A number of major gas producing states have recently revised regulations, with particular focus on emerging areas of concern, including disclosure of hydraulic fracturing chemicals, well construction and operation to prevent aquifer contamination, and management of waste from flowback and produced water. Similarly, many states and counties have some regulatory structures to address air quality issues based on each jurisdiction's individual circumstance. Some states and counties have adopted relatively stringent standards, while others have not. Some have rules proposed or are in the process of revising or implementing them. Others are considering rolling back existing regulations.[56] EPA's new federal air standards were drawn primarily from existing requirements found in the state codes of Colorado and Wyoming. For a comparison of EPA's new air standards for source categories in the crude oil and natural gas industry to those from selected states, see Table A-4 of this report.

Critics of federal oversight maintain that the states have long regulated oil and gas exploration and production and are best positioned to continue to do so as they have established governing structures and trained staff in place. They argue that a distant federal bureaucracy unfamiliar with local conditions is rarely the best entity to ensure environmental needs are balanced with economic growth and job creation. They claim that states can more readily address the regional and state-specific character of many crude oil and natural gas activities, including differences in geology, hydrology, climate, topography, industry characteristics, development history, state legal structures, population density, and local economics. Critics argue that federal rules add unnecessary and often repetitive requirements on the industry, which may increase project costs and delays with little added benefit. They claim federal rules would require increased federal resources and skilled staff to administer regulations for oil and gas development (or, as is sometimes the case, to administer federal requirements redundantly alongside existing state programs). They contend that a well-run state permitting and regulatory program can adjust more quickly, is better positioned to meet the challenges presented by constantly developing technologies, and can effectively administer rules across private, state, and federal lands.[57]

Others note that it has neither been the practice nor the intention of EPA to administer regulations for oil and gas development at the state level. However, seeing the patchwork mix of state and local requirements, they have called for the federal government to institute a consistent nationwide standard for emissions from the crude oil and natural gas sector. Proponents of federal oversight contend that many state laws and state agencies are still gaining experience with unconventional oil and natural gas development, and that industry is operating under insufficient and outdated rules.[58] They claim that many of the agencies that regulate development at the state level are underfunded and understaffed, are tasked with the dual purposes of developing the state's resources and protecting the state's environment, and thus, are caught between policing and promoting the industry. They suggest that state agencies and state regulators are often overwhelmed by oil and gas companies, which are generally national or international in scope, and start with the advantage of sheer size. They point to examples where state regulators have failed to seek large penalties for violations or track enforcement data, and cite statistics that show 40% of state drilling regulators have industry ties.[59] A 2012 report, released by a nongovernmental environmental organization, surveyed active oil and gas wells in six states and summarized the findings as follows: (1) over half (or close to 350,000 active wells in 2010) are operating with no independent inspections to determine whether they are in compliance with state rules; (2) when inspections do uncover rule violations, the violations often are not formally recorded; (3) when violations are recorded, they result in few penalties; (4) when penalties are assessed, they provide little incentive for companies to not offend again (noting that no state assessed annual fines that added up to the average value of a single gas well, about $2.9 million).[60]

Measurement of Emissions

- The new air standards are based on EPA's emissions estimates from the crude oil and natural gas sector. While emissions from certain activities and equipment in the sector lend themselves to credible estimates, others—specifically fugitive emissions from production activities such as hydraulically fractured well completions, flowback, and produced water ponds—are more difficult to evaluate, have fewer data available, and remain under considerable debate. Numerous sources have put forth a range of competing, and sometimes conflicting, reports over the past few years, but a comprehensive and consistent analysis of the types and quantities of fugitive emissions from the natural gas production sector does not currently exist. Until adequate and reliable measurements and analyses of emissions are undertaken, the benefits and cost of regulation or lack thereof may remain uncertain. In light of these considerations, Congress may examine ways in which to best facilitate more comprehensive and technically accurate emissions estimates.

By definition, "fugitive" emissions are those which are elusive and transitory. Thus, the single greatest difficulty in estimating emissions from natural gas systems is acquiring comprehensive and consistent measurement data. Currently, the most comprehensive study of emissions in the industry is more than a decade old, uses emissions factors and activity levels to calculate data, and focuses primarily on methane. More detailed greenhouse gas emissions from crude oil and natural gas production activities are scheduled to be reported by industry starting in 2012.[61] However, the national emissions inventories for criteria and hazardous air pollutants in the upstream sector of the industry are complicated by the fact that concentrations of these chemicals vary geographically across resource reservoirs. A few states have begun the process of acquiring emissions inventories from upstream production activities (e.g., Texas,[62] Pennsylvania,[63] and California,[64] as well as a national survey to be conducted by the University of Texas at Austin),[65] but few have presented or harmonized this information. There are also many examples of local emissions inventories, commissioned by a range of stakeholders—from regional and municipal agencies to community groups and academic institutions. None are fully consistent with shared measurement practices and each use different techniques for their data collection. It is for these reasons that the first recommendation in the report released by the U.S. Secretary of Energy Advisory Board Shale Gas Production Subcommittee is to "immediately launch projects to design and rapidly implement measurement systems to collect comprehensive methane and other emissions data."[66]

Currently, the primary source of information on emissions in the natural gas industry is a methane study published in 1996 by EPA and the Gas Research Institute (GRI).[67] At the time, the EPA/GRI study was conducted to assess the GHG emissions from various U.S. industrial sectors to assist in data analysis for the Intergovernmental Panel on Climate Change and emissions reporting for U.S. commitments to the United Nations Framework Convention on Climate Change.[68] The study focuses on 1992 (as a base year) and uses three primary methodologies to generate emissions factors from over 100 different sources within the industry.

The methods include (1) "component measurement," wherein emissions are measured directly from a large number of randomly selected pieces of equipment to determine an average emission factor for each type; (2) "tracer gas," wherein facility-wide emissions are calculated by releasing a tracer gas at a known and constant rate near the facility and measuring the downwind concentrations of the tracer and methane; and (3) "leak statistics," wherein emissions are measured for a large number of pipeline leaks to determine an average emissions rate per leak as a function of pipe material, age, operating pressure, and environmental characteristics. Total industry emissions are then estimated by multiplying these emissions factors by the activity levels for each system component (i.e., the number of wellheads, compressors, processing plants, miles of pipeline in operation, and other components) across the entire industry.

The EPA/GRI study estimates that the industry emitted 314 ± 105 billion standard cubic feet (bscf) of methane in 1992 (i.e., 127 ± 42 million metric tons of carbon dioxide equivalent [MMtCO2e], or 1.42% ± 0.47% of the industry's gross national production that year).[69] Roughly 60% of the industry's methane emissions are estimated to be from fugitive sources, about 30% from venting, and about 8% from combustion. Results of the EPA/GRI study, by source categories, are reported in Table 1.

Table 1. Methane Emissions in the U.S. Natural Gas Industry For the base year 1992

Source		Annual Methane Emissions (Bscf)	% of Total
Fugitive Emissions	Subtotal	195.3	62.1
Equipment Leaks			
Production Facilities		17.4	5.5
Gas Plants		24.4	7.8
Compressor Stations (transmission and storage)		67.5	21.5
Metering and Pressure Regulating Stations		31.8	10.1
Customer Meter Sets		5.8	1.8
Underground Pipeline Leaks (all segments)		48.4	15.4
Vented Emissions	Subtotal	94.2	30.0
Pneumatic Devices		45.7	14.6
Blow and Purge		30.2	9.6
Dehydrator Glycol Pumps		11.1	3.5
Dehydrator Vents		4.8	1.5
Chemical Injection Pumps		1.5	0.5
Other		0.9	0.3
Combusted Emissions	Subtotal	24.9	7.9
Compressor Exhaust		24.9	7.9
	Total	314.0	100.0

Source: David A. Kirchgessner et al., *Estimate of Methane Emissions from the U.S. Natural Gas Industry*, U.S. Environmental Protection Agency, Air Pollution Prevention and Control Division, Research Triangle Park, NC, 1996, p. 12.

Notes: Emissions from meter and pressure regulating stations result from both pneumatic and fugitive emissions. Since these components cannot be separated using the tracer measurement method, emissions are shown as fugitive by default.

EPA annually calculates emissions estimates for the industry using the methodology from the 1996 EPA/GRI study. Since its publication, activity data for some of the components in the system have been updated based on publicly available information. For other sources where annual activity data are not available or have not been reported by industry, EPA has developed a set of industry activity factor drivers to assist in modeling. While many of the emissions factors modeled by the EPA/GRI study were considered representative over the period of 1992 to the present, several factors have been re-calculated since publication. Most notably, emissions factors for gas well cleanups, condensate storage tanks, and centrifugal compressors have been revised due to new information. Emissions factors for gas well completions in unconventional resources with hydraulic fracturing—which were not industry practice at the time of the EPA/GRI study— have also been added to the inventory. With these revisions, EPA now estimates that the industry emitted 247.7 million metric tons of carbon dioxide equivalent (MMtCO2e) greenhouse gases in 2010, of which 215.4 MMtCO2e was methane (i.e., 532 bscf of methane, or 2.4% of the industry's gross national production that year).[70]

The emissions estimates EPA uses for hydraulically fractured well completions and re-completions come from data provided by industry sources at several EPA Natural Gas STAR technology transfer workshops between 2004 and 2007.[71] Using the reported data in its 2010 Background Technical Support Document, *Greenhouse Gas Emissions Reporting from the Petroleum and Natural Gas Industry*,[72] EPA creates separate categories for conventional and unconventional well completions, and increases its estimate of methane emissions from both categories from 0.02 metric tons per well completion to 0.71 metric tons per conventional well and 177 metric tons per unconventional well (or 37 thousand cubic feet [Mcf]/completion and 9,175 Mcf/completion respectively). EPA assumes a 3-10 day flowback period with an uncontrolled release of methane. Further, EPA assumes that 51% of the fugitive emissions from hydraulically fractured well completions are flared—and the rest vented—based upon state and local regulations for control devices that are currently in place.

EPA's methodology for estimating methane emissions from hydraulically fractured well completions has been criticized by a number of sources. A detailed critique of EPA's numbers can be found in IHS CERA's 2011 report, *Mismeasuring Methane: Estimating Greenhouse Gas Emissions from Upstream Natural Gas Development.*[73] CERA's main concerns involve the small sample size for emissions data and the unsupported assumptions on venting practices. Further, several industry sources have reported competing emissions estimates, including the American Petroleum Institute and the America's Natural Gas Alliance (which list total industry emissions at half of EPA's estimate due to a re-calculation of emissions factors and activity levels for liquids unloading[74] and re-fractured well completions),[75] URS (which estimates emissions of 765 Mcf of gas on a per well basis compared to EPA's 9,175 Mcf),[76] and Devon Energy Company (which reports data from eight of its operators demonstrating that flowback periods last on average only 3.5 days compared to EPA's estimated 3-10 days).[77] As with any self-selected and self-reported survey, it is difficult to determine how representative these samples are of overall industry practice.

As debate continues over the level of methane emissions from natural gas production activities, a full assessment of VOCs, HAPs, hydrogen sulfide, and other aromatic hydrocarbons becomes even more complicated, since the concentrations of these chemicals can vary greatly from reservoir to reservoir. Many observers note that the new air standards use methane emissions as a surrogate for VOCs and H2S/SO2 (e.g., EPA calculates cost effectiveness based on a national average of 3.7% VOCs by volume [18% by weight] for natural gas streams; however, many streams may produce little or no VOCs).[78] Thus, an accurate assessment of the levels of VOCs, SO2, and HAPs emissions from various production activities would require not only an accurate measurement of natural gas emissions but a full accounting of the component compositions of the gas across the geographic variability of the resource.

Impacts of Emissions

- The new air standards are based on EPA's expectations that the avoided emissions under the rules would result in improvements in air quality and reductions in health effects associated with exposure to HAPs, ozone, methane, and fine particulate matter. However, the relationship between air pollution from the crude oil and natural gas sector and its impacts on human health and the environment has never been fully quantified and assessed. By EPA's own admission during the rule proposal, the quantification of health benefits for the new standards could not be accomplished in a "defensible way" due to the "unavailability of data and the lack of published epidemiological studies correlating crude oil and natural gas production to respective health outcomes."[79]

  Some contend that this uncertainty argues against additional pollution controls. Others maintain that the relevant question for determining whether pollution controls are necessary is whether crude oil and natural gas systems contribute to a significant impact on an area's ability to attain air quality standards (NAAQS). In light of these considerations, Congress may wish to evaluate whether existing requirements are adequate to address concerns over the human health and environmental impacts of oil and gas activities, or, if different requirements are necessary.

Quantifying the extent to which crude oil and natural gas systems contribute to air pollution is a complicated task, due in part to the difficulties in modeling the direct and indirect impacts of emissions reductions from a single industrial sector on the greater environment. Although there are many different methodologies that can be used to assess emissions from the industry, each is burdened with unique disadvantages.[80] Thus, while science has demonstrated that emissions of VOCs and NO_x contribute to the generation of ground-level ozone (i.e., smog), emissions of various organic air toxics can pose a threat to human health, and emissions of methane produce ground-level ozone and force climate change, comprehensive epidemiological studies correlating the emissions from natural gas systems to specific long-range and cumulative health outcomes are virtually nonexistent. Such studies, however, are generally difficult, rare, and expensive to conduct, requiring data that are typically absent or inadequate for assessment (e.g., precise and accurate estimates of emissions, fate and transport, and exposure levels, as well as impacts data on relatively large populations of exposed individuals over long durations of time). More common, instead, are localized studies or anecdotal reports—by stakeholders who live and work near oil and gas operations—of general air quality issues such as haze, odor, or ill health. These studies are countered by industry and/or agency reports that present information on best practices, regulatory compliance, and a history of monitored and reported environmental stewardship.

Various stakeholders assert that the lack of published and peer-reviewed literature makes it challenging to scientifically assess the impacts of oil and gas drilling operations. The Centers for Disease Control and Prevention warns that the science on the impacts of natural gas drilling on health and the environment is "not yet clear," stating that there is "not enough information to say with certainty whether shale gas drilling poses a threat to public health," and that "more research is needed for us to understand public health impacts from natural gas drilling and new gas drilling technologies."[81] Similarly, a recent report by GAO[82] examines the available studies of air quality at shale gas development sites and finds that they are "generally anecdotal, short-term, and focused on a particular site or geographic location." Thus, they "do not provide the information needed to determine the overall cumulative effect that shale oil and gas activities have on air quality." GAO concludes that "the cumulative effect shale oil and gas activities have on air quality will be largely determined by the amount of development and the rate at which it occurs, and the ability to measure this will depend on the availability of accurate information on emission levels. However ... data on the severity or amount of pollutants released by oil and gas development, including the amount of fugitive emissions, are limited." It is for these reasons that the final recommendation in the report released by the U.S. Secretary of Energy Advisory Board Shale Gas Production Subcommittee is to advise federal, regional, state, and local jurisdictions "to place greater effort on examining [the] cumulative impacts" from "drilling and production operations, support infrastructure (pipelines, road networks, etc.) and related activities [which] can overwhelm ecosystems and communities."[83]

Some examples of studies on the human health and environmental impacts of oil and gas production are as follows:

Ozone

Many studies report that at each stage of natural gas production, VOCs—including BTEX, other hydrocarbons, and methane—can escape and mix with nitrogen oxides from the exhaust of diesel-fueled equipment, and, in the presence of sunlight, produce ground-level ozone (i.e., smog). Several areas of the country with heavy concentrations of drilling suffer from serious ozone problems.

For example, new oil and gas development has begun across sections of the Rocky Mountains that may be connected to the rise in ozone pollution across the area. A 2005 Western Governors' Association report finds that crude oil and natural gas production operations released more than 430,000 tons of VOCs in Colorado, New Mexico, Utah, Wyoming, and Montana in 2002. The report projects that operations in these states would more than double their VOCs emissions in 15 years, releasing more than 965,000 tons annually by 2018.[84] This release would equal the average amount of VOCs emitted annually from approximately 50,000 gas stations or by more than 25 million passenger cars, each driven 12,500 miles.[85] A 2008 analysis by the Colorado Department of Public Health and Environment concludes that smog forming emissions from Colorado's crude oil and natural gas operations exceed vehicle emissions for the entire state.[86] In 2009, the governor of Wyoming recommended that the state designate Wyoming's Upper Green River Basin as an ozone nonattainment area.[87] An extended assessment by the Wyoming Department of Environmental Quality finds the state's ozone pollution problems are "primarily due to local emissions from oil and gas ... development activities: drilling, production, storage, transport, and treating."[88]

Recently, northeastern Utah recorded unprecedented ozone levels in the Uintah Basin, with more than 68 exceedances of the federal health standard during the first three months of 2010, and over 24 exceedances during the winter of 2011.[89] The Bureau of Land Management identifies oil and gas activities in the region as the primary cause of the ozone pollution.[90] Further, a 2009 study concludes that numerous "Class I areas" in the Rocky Mountain region—a designation reserved for national parks, wilderness areas, and other such lands—are likely to be impacted by increased ozone pollution as a result of oil and gas development, including Mesa Verde National Park and Weminuche Wilderness Area in Colorado and San Pedro Parks Wilderness Area, Bandelier Wilderness Area, Pecos Wilderness Area, and Wheeler Peak Wilderness Area in New Mexico. These areas are all near concentrated oil and gas development in the San Juan Basin.[91] However, the evidence linking oil and gas development to high ozone levels in the region remains inconclusive. Researchers caution that many factors can contribute to the Rocky Mountains' ozone problems. For example, uniform snow cover reflects and concentrates sunlight, boosting ozone levels, as do temperature inversions that trap pollutants in mountain basins.[92] In support of this point, the interim findings of the *2012 Uintah Basin Winter Ozone & Air Quality Study,* released in August 2012, report that there were no exceedances of the eight-hour ozone standard for the basin in 2012, even though higher levels of ozone precursors and VOCs were measured in the region (the lack of sunlight-reflecting snowpack in 2012 is suggested as a possible reason).[93]

As another example, the Dallas-Fort Worth area in Texas is home to substantial oil and gas development. Of the nine counties surrounding the Dallas-Fort Worth area that EPA has designated as "nonattainment" for ozone, five contain significant oil and gas development.[94] A 2009 Southern Methodist University study finds summertime emissions of smog-forming pollutants from oil and gas operations in the Dallas-Fort Worth area exceed emissions from all motor vehicles in the area.[95] A 2012 study finds that emissions from natural gas compressor stations and flares contribute to significant amounts of ground-level ozone and formaldehyde in the Dallas-Fort Worth area, estimating that a single natural gas processing facility in the Barnett shale area could add as much as 3 parts per billion (ppb) to the hourly average ambient ozone.[96] Still, the evidence linking oil and gas development to high ozone levels in the region remains inconclusive. In 2009 and 2010, the Texas Commission on Environmental Quality (TCEQ) conducted several large, in-depth surveys of air quality in the six counties surrounding Fort Worth. The TCEQ study identifies a 15% drop in the 8-hour ozone design value, despite a

10-fold increase in natural gas production in the region over the past decade. TCEQ attributes at least part of this reduction to the state's NO_x control strategies as well as the prevailing winds.[97]

Air Toxics

Other studies (e.g., epidemiological studies) focus on the human health impacts of hazardous air pollutants (HAPs) known or suspected to be released by crude oil and natural gas operations. A 2011 survey of existing literature on the topic reported that of the known chemicals used and/or found in natural gas operations, approximately 37% of the chemicals are volatile and may become airborne. The study further noted that if exposures exceed certain levels, over 89% of these chemicals can harm the eyes, skin, sensory organs, respiratory tract, gastrointestinal tract, or liver; 81% can cause harm to the brain and nervous system; 71% can harm the cardiovascular system and blood; and 66% can harm the kidneys. Overall, the hazardous air pollutants produce a profile that displays a higher frequency of health effects than the water soluble chemicals. In addition, because they vaporize, not only can they be inhaled, but they can be ingested or absorbed through the skin, increasing the chance of exposures.[98]

Several studies report that sources of HAPs in the oil and gas industry may be numerous. Data from the State of Colorado suggest that there may be more than 26 individual sources of HAPs in the oil and gas sector, including "venting, dehydration, gas processing, compression, leaks from equipment (fugitive emissions), open-pit waste ponds, and land application of volatile wastes."[99] A study of HAPs emissions from natural gas related sources within the city of Fort Worth, TX, documents the following: HAPs emissions of 0.02 to 2 tons per year (tpy) from well pads (including emissions from equipment leaks, produced water and condensate storage and loading, and lift compressors); HAPs emissions of 0.9 to 8.8 tpy from well pads with compressors; HAPs emissions of 10 to 25 tpy from compressor stations (including emissions from combustion at the compressor engines or turbines, equipment leaks, storage tanks, glycol dehydrators, flares, and condensate and/or wastewater loading); HAPs emissions of 47 tpy on average from processing facilities (including emissions from equipment leaks, storage tanks, separator vents, glycol dehydrators, flares, condensate and wastewater loading, compressors, amine treatment and sulfur recovery units); and HAPs emissions of 0.4 tpy on average from saltwater treatment facilities.[100]

Although oil and gas development generally has not occurred in densely populated areas, this is not always the case, particularly with respect to recent expansion in unconventional resources. Several local, state, and national health agencies have expressed concerns about the health impacts of HAPs emissions from oil and gas facilities, including the Center for Disease Control and Prevention (CDC),[101] the Agency for Toxic Substances and Disease Registry (ATSDR),[102] the Association of Occupational and Environmental Clinics (AOEC) and the Pediatric Environmental Health Specialty Unit (PEHSU),[103] the Colorado School of Public Health,[104] the Town of Dish, Texas,[105] the City of Fort Worth, Texas,[106] and the City of Houston, Texas.[107] In particular, the ATSDR investigation was spurred by community health complaints such as dizziness, nausea, respiratory problems, and eye and skin irritation to more severe concerns including cancer. The investigation identifies elevated cancer risk at one site and recommends further investigation into HAPs emissions and risks at all the sites. Similarly, the Colorado School of Public Health identifies air pollution from the crude oil and natural gas activities as contributing to acute and chronic health problems for those living near natural gas drilling sites, stating that "exposures to air pollutants during well completion activities present the greatest potential for health effects."[108] The CSPH analysis, based on three years of monitoring, finds a number of air toxics near the wells including benzene, toluene, ethylbenzene, and xylene.

However, critics have taken issue with the Colorado School of Public Health and other studies for a number of reasons including the use of (1) out of date data, (2) unrealistic estimates of emissions from sources, (3) overly conservative risk assessments when viewed in context, and (4) poor study design and input assumptions (e.g., the use of ambient collection methodology in a region populated with other emission sources such as roads, highways, and other industries).[109] In counterpoint, other studies have reported little or no health effects from the sector. For example, in 2012, the Houston-based Plains Exploration and Production Company (PXP) released a report that said hydraulic fracturing operations at the Inglewood oil field in the Baldwin Hills area in Los Angeles County posed "no public health or environmental threats."[110] Similarly, in a review of health studies conducted by the Energy Institute at the University of Texas at Austin and published in February 2012, the authors conclude that "none of the studies reviewed ... showed a clear link between shale gas activities and documented adverse health effects, [and] that the gas industry has been using hydraulic fracturing for over 50 years, but the studies examined ... did not find any direct evidence for health impacts on workers in the industry or the public living near oil and gas industry activity."[111] Further, many industry sources have argued that the standards promulgated under the 1999 Crude Oil and Natural Gas Production NESHAPs have demonstrated protection of public health with an ample margin of safety.[112] Critics of EPA's regulatory requirements for air toxics have also expressed concern with the agency's decision to make significant and substantive changes to the agency's residual risk procedures, changes—which they assert—result in risk estimates between 100 and 1,000 times higher than actual. The contested procedures include (1) the consideration of risk from the total facility, (2) the consideration of risk across selected social, demographic, and economic groups within the population living near the facility, (3) the consideration of the hypothetical risk associated with the level of emissions allowed by the MACT standard, and (4) unreasonable assumptions concerning exposure points, times, and durations.

Greenhouse Gases

Finally, other studies focus on the impact of greenhouse gas emissions from crude oil and natural gas operations. Many of these studies advocate for the increased production and use of natural gas as a substitute for other fossil fuel resources because it is domestically available, economically recoverable, and considered a potential "bridge" fuel to a less polluting and lower GHG-intensive economy. However, when fugitive emissions of methane from all sectors of the industry are factored into the total life-cycle assessment of the resource,[113] the advantage of natural gas over other fossil fuels is less clear. EPA's *Inventory of U.S. Greenhouse Gas Emissions and Sinks: 1990-2010* (published April 15, 2012) estimates natural gas systems emitted 215.4 million metric tons of carbon dioxide equivalent (MMtCO2e) of methane in 2010, a 14% increase over 1990 emissions, and 32.3 MMtCO2e of non-combustion CO2 in 2010, a 14% decrease over 1990 emissions. EPA cites improvements in management practices and technology, along with the replacement of older equipment, as helping to stabilize emissions of CO2. However, an increase in activities and production wells contributed to the higher levels of methane.[114] In 2010, natural gas systems were cited as being the single largest contributor to U.S. anthropogenic (i.e., man-made) methane emissions, representing 32.3% of the total methane emissions from all domestic sources and accounting for about 3% of all GHG emissions in the United States. Some recent studies (outlined below) argue that expanding the production and use of natural gas could actually exacerbate rather than help mitigate global warming because methane's global warming potential (GWP)[115] is as much as 20—if not 50—times more potent than carbon dioxide, depending upon the time interval used to express warming impacts. These studies suggest that if significant amounts of methane were to leak during the production of natural gas, the climate forcing of methane could cancel out or even outweigh any benefit gained from switching away from oil or coal combustion processes.

Many of the published and publicly available life-cycle GHG emissions assessments of natural gas rely on the 1996 EPA/GRI emissions factors for methane leakage in the industry. EPA's recently updated emissions estimates suggest that leakage from natural gas systems is approximately 2.4% of gross national production.[116] EPA has neither calculated nor stated whether these revisions negate the apparent advantage natural gas has over oil or coal when it comes to climate change. But some researchers, most notably Robert Howarth of Cornell University, have argued that they would. Howarth estimates that emissions from natural gas leakage are likely to be in the range of 3.6% to 7.9% of gross national production over the entire natural gas life-cycle (which includes production, processing, transmission, storage, and distribution). He estimates leakage at the wellhead during both well completion and production to be between 2.2% and 3.8%.[117] At these rates, "the footprint of shale gas is at least 20% greater [than coal] and perhaps more than twice as great on the 20-year horizon and is comparable when compared over 100 years." Others, however, argue that Howarth's analysis is flawed, claiming that he significantly overestimates the fugitive emissions associated with unconventional gas extraction, undervalues the contribution of "green technologies" (e.g., REC) to reducing these emissions, bases the comparison between gas and coal on heat rather than electricity generation (almost the sole use of coal), and assumes a time interval (i.e., 20 years) over which to compute the relative climate impact of gas compared to coal, thus failing to capture the contrast between the long residence time of CO2 and the short residence time of methane in the atmosphere.[118]

The controversy has led to a number of life-cycle GHG assessments comparing natural gas to coal. Most use the emissions measurement data from the 1996 EPA/GRI study as a starting point for their assumptions. For example: Hultman (2011)[119] estimates that for electricity generation, the GHG impacts of shale gas are 11% higher than those of conventional gas, and about half that of coal for standard assumptions. Fulton (2011)[120] concludes that U.S. gas-fired electricity generation emits 47% less GHG than coal using the IPCC's 100-year global warming potential index for methane, instead of the 20-year index used by Howarth. Jiang (2011)[121] estimates that emissions from coal-fired electric generation are so much greater than gas-combustion plants that a fuel switch would reduce climate-forcing emissions as long as leakage in a natural gas system is less than 14% (using the IPCC 100-year GWP) or 7% (using the IPCC 20-year GWP). Alvarez (2012),[122] however, in an assessment similar to Jiang's, concludes that leakage in the natural gas system would only need to be 3.2% before the overall lifecycle emissions estimates for natural gas electricity generation surpassed those of coal. Further, powering vehicles with natural gas rather than gasoline would not strike the same balance in the near term unless gas producers bring their leakage rate down to a maximum of 1.6%; and replacing diesel with natural gas would cause greater near-term climate impacts unless leak rates were under 1.0%.

Estimating the GHG Footprint of Unconventional Natural Gas

Life-cycle assessment (LCA) is an analytic method used for evaluating and comparing the environmental impacts of various products (in this case, the climate change implications of hydrocarbon resources). LCA can be used in this way to identify, quantify, and track emissions of carbon dioxide and other GHG emissions arising from the development of these hydrocarbon resources, and to express them in a single, universal metric of carbon dioxide equivalent (CO2e) GHG emissions per unit of fuel or fuel use. This figure is commonly referred to as the "emissions intensity" of a fuel. The results of an LCA can be used to evaluate the GHG emissions intensity of various stages of the fuel's life-cycle, as well as to compare the emissions intensity of one type of fuel or method of production to another.

As researchers point out: "Gas produced from unconventional wells has roughly the same methane content as that produced from conventional wells and therefore combustion can be assumed to yield the same climate effect. However, extraction techniques for unconventional gas differ from those used for conventional gas, and figures on well-lifecycle methane emissions have not been comprehensively established. Calculating the GHG footprint of unconventional gas requires three steps and associated assumptions. First, emissions of GHG from the production process, leaked methane and CO2, must be estimated. Second, these numbers must be converted to a common GHG metric such as CO2-equivalent (CO2e). Third, because electricity generation technologies vary greatly in their combustion efficiencies, the emissions attributable to a kilogram or GJ of fuel are more appropriately compared on the basis of electricity delivered to the end-user—i.e. on a per kWh basis."[123]

Different published studies have approached these three inputs differently, accounting for the wide spectrum of estimates for natural gas's GHG footprint in comparison to coal, oil, or other fuel sources.

Other published studies have used different methodologies for the calculation of leakage (e.g., satellite observations, ambient measurements, and dispersion modeling). A National Oceanographic and Atmospheric Administration (NOAA) study analyzes daily air samples collected at the Boulder Atmospheric Observatory in Weld County in northeastern Colorado and concludes that fugitive natural gas emissions from drilling operations range from 2.3% to 7.7% of industry's gross annual production.[124] The study uses an extensive data set of ambient concentrations of methane and related hydrocarbons in the vicinity of oil and gas operations in the region, along with the known emissions profiles for these gases from oil and gas operations, to infer the emissions from the industry.

Cost Benefit Analysis of Federal Standards

- The new air standards require producers to capture about 90% of the natural gas that escapes into the atmosphere as a result of production using hydraulic fracturing. Methane and other emissions found in natural gas streams are a significant source of pollution when they escape into the atmosphere. Methane, however, is also the main product of—and thus a main source of revenue for— the industry. Due to this unique linkage, control over leakage has the potential to translate into economic benefits, as producers can offset compliance costs with the value of natural gas and condensate recovered and sold at market. EPA estimates that the equipment and the activities required to comply with the new standards will cost producers about $170 million a year, but calculates that incorporating the sale of recovered products into the cost will result in an estimated net gain of about $11 million to $19 million a year. The industry counters with compliance cost estimates at more than $2.5 billion annually. Third parties, such as Bloomberg Government, project a net cost between $316 million and $511 million, or, approximately 1% of industry's annual revenue. All estimates are based on assumptions regarding the quantity of captured emissions, the cost of capital equipment, and the market price for natural gas. Thus, until adequate and reliable measurements and analyses of emissions are undertaken, the benefits and cost of regulation (or lack thereof) may remain uncertain. In light of these considerations, Congress may opt to examine ways in which to best align the costs of requirements with sustainable resource expansion, industry growth, and job creation.

EPA estimates the new air standards would yield a cost savings of $11 million to $19 million in 2015, as pollution abatement costs would be offset by the value of natural gas and condensate recovered and sold at market. EPA's annual cost estimates include (1) cost of capital, (2) operating and maintenance costs, (3) cost of monitoring, inspection, recordkeeping and reporting (MIRR), and (4) associated product recovery credits. All costs are reported in 2008 dollars, and calculated as follows: The total estimated net annual cost to industry to comply with the final amendments in the Crude Oil and Natural Gas Production NESHAPs is approximately $3.3 million. The total estimated net annual cost in the Natural Gas Transmission and Storage NESHAPs is approximately $180,000. The total estimated net annual cost to industry to comply with the final NSPS is approximately $170 million; and, when revenues from additional product recovery are considered, the final NSPS is estimated to result in a net annual engineering cost savings of approximately $15 million. EPA assumes that producers will be paid $4/Mcf for the recovered gas at the wellhead and $70 per barrel for recovered condensate, with about 43 million Mcf (i.e., 43 billion cubic feet) of natural gas and 160,000 barrels of condensate recovered by control requirements in 2015. Further, EPA estimates that the operation cost of 3-10 days of hydraulic fracturing with reduced emissions completions would be $700 to $6,500 per day with an additional $3,500 per well for flare equipment, for an average incremental cost of using the abatement technology of $33,237 per completion. EPA annualizes the cost using a 7% discount rate.

The American Petroleum Institute (API), in its comments to EPA on the proposed regulation, included a consultant's report that projects compliance costs of more than $2.5 billion annually after accounting for incremental sales of the fuel.[125] API estimates the cost of flaring emissions to be $90,000 per well, and the operation cost of using (i.e., renting and setting up) reduced emissions completions to be $180,000 per completion. API reports that only 300 REC sets are currently in existence, with the capacity to produce only 4,000 wells a year. Working with EPA's projection that there would be 25,000 new or modified hydraulically fractured wells completed annually, API argues that the equipment prescribed to conduct reduced emission completions would not be available in time to comply with the rule schedule. Further, API contends that EPA's cost analyses are based on "average model facilities" that do not represent all equipment and compliance costs and fail to identify when the controls are no longer economic. At the time of the proposed rule, API's consultant claimed that the new air standards could create "a significant slowdown in unconventional resource development ... resulting in less reserve additions, less production, lower royalties to the Federal government and private landowners, and lower severance tax payments to state governments. The delays in drilling result in delays in production, which result in the delays in the economic benefits associated with that production." Other industry commentators have gone further by stating that the full burden of regulations would fall squarely upon the producers, because economic incentives dictate that producers are currently capturing and selling all economically recoverable natural gas. EPA's rule would "divert investments from capital and energy development into regulatory compliance efforts, and impose onerous notification, record keeping, monitoring, reporting, and performance testing requirements that industry will necessarily incur costs to keep up with."[126]

In the final *Regulatory Impact Assessment* issued in April 2012, EPA examines the compliance costs for slightly more than 11,000 hydraulically fractured wells that would be drilled in unconventional formations, and nearly 1,500 that would be re-fractured to stimulate production. Of those, approximately 6,600 are assumed to be using reduced emissions completions voluntarily or under state/local regulations, and an additional 12% would be classified as exploratory, delineation, or low-pressure wells exempt from the new standards, leaving about 4,600 that would be compelled to use REC as a result of the new federal regulation.[127] Industry, however, claims that most large producers in the natural gas sector are already using REC where feasible, and thus EPA's recovery estimates are optimistic. API has stated that most large producers are currently using REC "where it makes economic sense," but that many smaller and independent producers may not yet be using the technique. Chesapeake Energy, the number two U.S. gas producer, has said it limits emissions on "a very high percentage" of its wells. Fourth-ranked Devon Energy has stated it uses REC for 91% of its wells. A survey of nine API members working in 29 drilling areas reported their members doing REC on 70% of their wells. America's Natural Gas Alliance (ANGA) submitted a survey to EPA of eight unidentified production companies that report using REC on 93% of their wells in 2011.[128] As with any self-selected and self-reported survey, it is difficult to determine how representative these samples are of overall industry practice.

A comparative report by Bloomberg Government—using slightly different values to analyze the available data on affected wells, compliance costs, products recovered, and market prices— calculates that the costs would fall somewhere in between both EPA's and API's projections. In their analysis, the regulation would generate an initial $125 million in spending on new equipment and $383 million in annual rental and service fees beginning in 2015. Factoring in product recovery, the total net costs of compliance would fall between $316 million and $511 million per year depending upon the market price of natural gas—or, more than twice the amount estimated by EPA and less than half that projected by API. This increase would account for approximately 0.5% to 0.7% of the industry's 2011 gas sales, which they determined "may cause a slight reduction in drilling activity."[129] Bloomberg Government noted that EPA's analysis assumed natural gas prices 40% above levels at the time and estimates that prices would need to rise to $10/Mcf to make the value of recovered products equal to the compliance costs.

Covered Sources and Pollutants

- The new air standards primarily focus on the upstream sectors of the oil and gas industry and cover only some of the pollutants and only some of the potential sources of emissions. The standards regulate emissions of volatile organic compounds from some, but not all, of the equipment and activities at on-shore natural gas well sites, gathering and boosting stations, and processing plants. Similarly, the standards regulate emissions of sulfur dioxide from sweetening units at some on-shore natural gas processing plants as well as hazardous air pollutants from some dehydration units and storage facilities in the sector. Many pollutants and sources remain uncovered by any standard, and some critics point specifically to methane emissions from the midstream and downstream sectors, as well as hydrogen sulfide, as the most significant omissions. In light of these considerations, Congress may decide to re-examine ways in which oversight activities address the most significant pollutants and point sources.

While the new air standards for crude oil and natural gas systems are more detailed and comprehensive than previous standards, several pollutants and sources in the sector remain uncovered.

The new standards do not directly cover emissions of the following pollutants in the sector: methane, nitrogen oxides, particulate matter, and hydrogen sulfide (although reductions in some of these pollutants may occur as a co-benefit of VOCs and SO_2 reductions). Many observers have noted, however, that the new standards, as written, use natural gas emissions as a surrogate for VOCs and H_2S/SO_2 emissions. Some argue that basing standards on the volume of natural gas emissions as opposed to the content of VOCs in the gas does not adequately account for the geographic variability of VOCs within the resource (e.g., EPA calculated cost effectiveness based on a national average of 3.7% VOCs by volume [18% by weight] for natural gas streams; however, many streams may produce little or no VOCs). They maintain that requiring standards for natural gas emissions as opposed to VOCs content essentially regulates the industry for methane as opposed to VOCs. From this perspective, requiring an operational standard (e.g., reduced emissions completions) instead of a performance standard (e.g., a VOCs threshold) may impose unnecessary compliance costs in some instances.[130] EPA did exempt many activities and pieces of equipment in the transmission, storage, and distribution sectors of the industry from the VOCs standards due to the smaller volume of VOCs in natural gas streams after the processing stage.

Further, the new standards do not cover emissions from the following sources in the sector: all oil wells; all off-shore sources; all coal-bed methane production facilities; all field engines, drilling rig engines, and turbines; well-head and transmission and storage segment compressors; wellhead activities such as liquids unloading; all heater-treaters; all pneumatic devices other than controllers; storage vessels such as skid-mounted, mobile, well cellars, sumps, and produced water ponds; and LDAR for non-processing plant facilities. Further, the new standards do not cover VOCs or SO_2 emissions from existing sources, unless they are classified as HAPs. Finally, the new standards assume non-gas-driven controllers cannot replace gas-driven controllers as "best system of emission reduction" (BSER) for regulatory purposes; low bleed controllers cannot replace high-bleed controllers as BSER; centrifugal compressors cannot replace reciprocating compressors as BSER; and vapor recovery units cannot replace combustion devices as BSER. Many of these sources may still emit significant quantities of pollutants; however, EPA has determined that standards on these sources are either technically or economically unfeasible.[131]

Major Source Aggregation

- The new air standards exempt well completions, pneumatic controllers, compressors, and storage vessels from Clean Air Act

> (CAA) Title V permitting obligations (i.e., from "major source" determinations for air pollution). Viewed at the component level, these smaller "emissions units" at natural gas production facilities may not generate enough pollution on their own to be classified as major sources. However, it may be possible that an entire natural gas operation (e.g., a well-site, a field, or a station) is a major source (i.e., one that emits typically 10-250 tons per year, depending upon the pollutant). Determining which equipment and activities should be grouped together, or "aggregated," in the crude oil and natural gas production sector for permitting purposes remains a contested issue for EPA, the courts, and the respective state agencies. In light of these considerations, Congress may monitor and assess EPA's plans to formulate the requirements for major source categories in upstream oil and gas activities.

The past few decades have seen several developments in the evaluation of crude oil and natural gas facilities at multiple locations for possible aggregation into a single source for permitting purposes. The ability of EPA and the state permitting authorities to aggregate multiple operations into a single major source permit is founded upon the definition of "stationary source" within the CAA. The CAA defines a "stationary source" as "any building, structure, facility, or installation which emits or may emit any air pollutant."[132] *Alabama Power Co. v. Costle*[133] established boundaries on the scope of a source such that "(1) it must carry out reasonably the purposes of New Source Review/Prevention of Significant Deterioration (NSR/PSD); (2) it must approximate a common sense notion of 'plant'; and (3) it must avoid aggregating pollutant-emitting activities that as a group would not fit within the ordinary meaning of 'building,' 'structure,' 'facility,' or 'installation.'"[134] In response, in the 1980 revisions to the PSD regulations, EPA clarified that emissions from operations may be aggregated and considered a single major source for PSD permitting if they meet each of the following three criteria: (1) the sources are located on one or more "contiguous or adjacent" properties, (2) the sources are under common control of the same person (or persons under common control), and (3) the sources belong to a single major industrial grouping (same two digit major Standard Industrial Classification (SIC) code). Only if all three criteria are met will the CAA permitting authority aggregate the operations into a single NSR/PSD permit. After the 1990 CAA Amendments created the Title V Operating Permit Program, this three-factor analysis was extended to Title V major source permitting.

The source definition established by EPA was intended to aggregate only "major projects that would cause air quality deterioration" but "avoid review of projects that would not increase deterioration significantly."[135] EPA has recently addressed the issue of CAA source determinations in the oil and gas industry in a 2009 guidance document from the EPA Office of Air and Radiation (the "McCarthy Memo").[136] The McCarthy Memo withdrew earlier guidance from EPA which concluded that the three prong aggregation analysis for oil and gas activities should begin by looking at and focusing most heavily on the proximity of the surface locations. This emphasis on proximity may have been a result of previous actions by EPA that interpreted "contiguous and adjacent" as meaning "functionally interdependent" (e.g., sources connected by pipelines, conveyors, roads, and other means by which materials and products or intermediate products are transferred between them). The McCarthy Memo attempted to negotiate a path between the broad mandate of "functional interdependence" and the more narrow use of "proximity." It recognized that source determinations in the oil and gas industry continue to be complex, and re-emphasized that the regulations list all three criteria to be used in the analysis. It then acknowledged that there would be cases in which proximity is the "overwhelming factor," but the agency is not going to pre-judge that by using a simplified approach, and that "reasoned decision-making" of each of the relevant factors needs to occur on a case-by-case basis.

EPA's three prong aggregation analysis for oil and gas activities was recently challenged in the case of *Summit Petroleum Corp. v. EPA* before the U.S. Court of Appeals for the 6th Circuit.[137] In an August 7, 2012, decision, the 6th Circuit rejected 2-1 the agency's "functional interrelationship" analysis used to support its definition of "adjacency," vacating EPA's determination that a Michigan natural gas plant and its production wells constitute a single major source.

Further, on October 29, the court issued an order denying EPA's motion to rehear the case.

The denial cements the appellate court's ruling, remanding the issue back to EPA, and ordering the agency to conduct a new analysis of the Michigan facilities using physical proximity as the sole basis for determining adjacency. In response, EPA has stated that relying on physical proximity alone may lead both to "absurd results" and increased regulatory burdens under the NSR and PSD programs.[138] EPA has not stated its intentions in response to the denial, but observers point out that an appeal to the U.S. Supreme Court is unlikely because there is no circuit split.

Furthermore, with respect to hazardous air pollutants, it should be noted that in Section 112(n)(4) of the CAA, Congress specifically exempted upstream oil and gas operations from aggregation in several ways. First, with respect to the "major source" category, Section 112(n)(4) of the CAA provides that, notwithstanding the general definition of "major source,"

> emissions from any oil or gas exploration or production well (with its associated equipment) and emissions from any pipeline compressor or pump station shall not be aggregated with emissions from other similar units, whether or not such units are in a contiguous area or under common control, to determine whether such units or stations are major sources, and in the case of any oil or gas exploration or production well (with its associated equipment), such emissions shall not be aggregated for any purpose under this section.

Second, with respect to the "area source" category, Section 112(n)(4) provides that

> the Administrator shall not list oil and gas production wells (with its associated equipment) as an area source category under subsection (c), except that the Administrator may establish an area source category for oil and gas production wells located in any metropolitan statistical area or consolidated metropolitan statistical area with a population in excess of 1 million, if the Administrator determines that emissions of hazardous air pollutants from such wells present more than a negligible risk of adverse effects to public health.

At the time, the 101st Congress (1990) found that

> oil and gas wells, and associated equipment and gas processing, have generally very low emissions of air toxics. Furthermore, these operations are typically located in remote areas, with wells and equipment widely dispersed geographically, rather than concentrated in a single area. For these reasons, it is very unlikely that oil and gas sources would present a significant risk to human health and it is not expected that this source category would need to be a listed category designated for regulation.[139]

In response, EPA wrote in the 1999 preamble that the definition of facility should "lead to an aggregation of emissions in major source determinations that is reasonable, consistent with the intent of the Act, and easily implementable."

Consequently, EPA determined it was not appropriate to aggregate crude oil and natural gas facilities at that time.[140]

Recent Legislation

- Recent legislation to amend the Clean Air Act to address emissions from crude oil and natural gas production activities includes:

The BREATHE Act

On March 17, 2011, during the 112th Congress, Representative Jared Polis (D-CO-2), Representative Maurice Hinchey (D-NY-22), and 20 other co-sponsors introduced H.R. 1204, a bill "to amend the Clean Air Act to eliminate the exemption for aggregation of emissions from oil and gas development sources, and for other purposes." Dubbed the Bringing Reductions to Energy's Airborne Toxic Health Effects Act, or BREATHE Act, the proposed bill would amend the CAA to (1) include hydrogen sulfide in the list of hazardous air pollutants; (2) repeal the prohibition on aggregating emissions from any oil or gas exploration or production well and emissions from any pipeline compressor or pump station with emissions from other similar units to determine whether such units or stations are major sources of hazardous air pollutants; (3) repeal the prohibition on aggregating emissions from any oil or gas exploration or production well for any purpose relating to hazardous air pollutant emission standards; and (4) repeal the prohibition against the Administrator listing oil and gas production wells as an area source category of hazardous air pollutants. The bill was referred to the House Subcommittee on Energy and Power on March 28, 2011.

The Empower States Act

On September 19, 2012, during the 112th Congress, Senator John Hoeven (R-ND) and co-sponsor Lisa Murkowski (R-AK) introduced S. 3573, a bill "to recognize the primacy of States, provide for the consideration of the economic impact of additional regulations, and provide for standards and requirements relating to certain guidelines and regulations relating to health and the environment." The proposed bill would "ensure states retain the right to regulate hydraulic fracturing" and blocks the EPA from taking any action against a state, an industry, or an individual unless it can demonstrate "imminent and substantial danger to the public health or environment exists and that the state has failed to take corrective action." It further would require any federal department or agency that plans to draft regulations relating to oil and gas development to consult with the state, submit an "Energy and Economic Impact Statement," and show that the state does not have existing alternatives to the new regulation. The bill was referred to the Senate Committee on Environment and Public Works on September 19, 2012.

CONCLUSION

Federal oversight of the natural gas industry remains controversial. Many have advocated for the increased production and use of natural gas in the United States for economic, national security, and environmental reasons. They argue that increased oversight may slow development and curtail the various economic and environmental co-benefits of increased domestic production. Traditionally, natural gas has been seen as the cleanest burning fossil fuel, with fewer emissions of carbon dioxide, nitrogen oxide, sulfur dioxide, particulate matter, and mercury than its hydrocarbon rivals (e.g., coal and oil) on a per unit of energy basis. Many have looked to it as a "bridge" fuel to a less polluting and lower GHG-intensive economy. However, the recent expansion in natural gas production in the United States has given rise to a new set of concerns regarding the human health and environmental impacts of natural gas. The effectiveness of current regulatory oversight is one of these concerns.

Unlike the debate over groundwater contamination or induced seismicity—where questions remain as to whether or not operations contribute significantly to these impacts—there is little question that natural gas systems emit air pollutants. Air pollutants—including volatile organic compounds (a ground-level ozone, or smog, precursor), sulfur dioxide, methane (a greenhouse gas), and various hazardous air pollutants—are released from activities and equipment throughout the natural gas industry. The recent increase in hydraulic fracturing operations, as well as the encroachment on more heavily populated areas, has only served to heighten the concern over emissions from the industry. These concerns are reflected in such findings as (1) significant increases in the levels of VOCs and/or ozone in some rural areas near new natural gas development, (2) increased incidences of odors, respiratory ailments, and other health issues in some populated areas near new natural gas development, and (3) increased concern over the climate impacts of fugitive methane released from natural gas development.

To address air quality and other environmental concerns, the development of natural gas in the United States is regulated under a complex set of local, state, and federal laws. Currently, state and local authorities are responsible for virtually all of the day-to-day oversight of natural gas development, and many states and counties have regulatory structures in place to address air quality issues based on local needs. Additionally, the U.S. Environmental Protection Agency has promulgated minimum standards for VOCs, SO2, and HAPs in some sectors of the industry for over a decade. Recently, in response to a consent decree issued by the U.S. Court of Appeals, D.C. Circuit, EPA promulgated a new set of air standards to regulate emissions of VOCs, SO2, and HAPs from many source categories that had never before been covered by federal oversight. The new standards focus primarily on the upstream sector of the industry, and were drawn, in part, from existing requirements found in the state codes of Colorado and Wyoming. They regulate (1) VOCs emissions from gas wells, centrifugal compressors, reciprocating compressors, pneumatic controllers, storage vessels, and leaking components at onshore natural gas processing plants, (2) SO2 emissions from onshore natural gas processing plants, and (3) HAPs emissions from dehydrators and storage vessels. The most significant provision in the new standards is the requirement to use "reduced emissions completions" on all hydraulically fractured or refractured on-shore natural gas wells (i.e., the requirement to capture and use the gas flowback that is otherwise vented or flared).

Many believe that standards for the natural gas industry which are similar to those that have been promulgated by Colorado, Wyoming, and the federal government could significantly reduce VOCs, SO2, and HAPs emissions. Some argue that the cost of compliance to these air standards could reduce industry margins, thereby reducing economic interests to invest and slowing production activities.

Others are concerned that some pollutants and some emissions sources remain uncovered by any standard, including methane, hydrogen sulfide, oil wells, off-shore wells, conventional gas wells, and most operations downstream of the processing plant.

Debate over the costs of compliance and the proper regulatory institutions (i.e., local, state, or federal) continues. Despite these concerns, adequate and reliable measurements of emissions from the natural gas industry do not currently exist. This is due to many factors, including costs and technical uncertainties. But, until these measurements are undertaken and analyzed, the benefits and costs of regulation or lack thereof may remain uncertain.

APPENDIX A. FEDERAL AIR STANDARDS FOR CRUDE OIL AND NATURAL GAS SYSTEMS

Table A-1. New Source Performance Standards for Volatile Organic Compounds in Crude Oil and Natural Gas Systems

Source Sector	Affected Facility	Prior Rules	New Rules	Compliance Dates	Comment
		40 C.F.R. Part 60, Subpart KKK	40 C.F.R. Part 60, Subpart OOOO	77 FR 49489, August 16, 2012	New or Modified Facilities Defined as Beginning Construction After August 23, 2011
Natural Gas Well Sites	New Hydraulically Fractured Well Completions	Not covered.	Exploratory wells: operators must use completion combustion device (e.g., flaring) unless hazardous or prohibited by state or local regulations.	October 15, 2012.	Exploratory well defined as a well outside known fields or the first well drilled in an oil or gas field where no other oil and gas production exists.
			Delineation wells: operators must use completion combustion device (e.g., flaring) unless hazardous or prohibited by state or local regulations.	October 15, 2012.	Delineation well defined as a well drilled in order to determine the boundary of a field or producing reservoir.
			Hydraulically fractured low-pressure wells: operators must use completion combustion device (e.g., flaring) unless hazardous or prohibited by state or local regulations.	October 15, 2012.	Low pressure well defined by a formula based on well depth and pressure. (A well with reservoir pressure and vertical well depth such that 0.445 times the reservoir pressure (in psia) minus 0.038 times the vertical well depth (in feet)

Source Sector	Affected Facility	Prior Rules	New Rules	Compliance Dates	Comment
					minus 67.578 psia is less than the flow line pressure at the sales meter.) EPA estimates that this exclusion for low pressure gas wells would cover 10% of all natural gas wells and, specifically, 87% of coal-bed methane wells.
			All other hydraulically fractured wells, prior to January 1, 2015: operators must use either completion combustion device (e.g., flaring) or reduced emissions completion (i.e., "green completion").	Prior to January 1, 2015.	Completion combustion device defined as any ignition device, installed horizontally or vertically, used in exploration and production operations to combust otherwise vented emissions from completions.
			All other hydraulically fractured wells, after January 1, 2015: operators must use reduced emissions completion (i.e., "green completion").	On or after January 1, 2015.	Reduced emissions completion defined as a well completion following fracturing or refracturing where gas flowback that is otherwise vented is captured, cleaned, and routed to the flow line or collection system, re-injected into the well or another well, used as an on-site fuel source, or used for other useful purpose that a purchased fuel or raw material would serve, with no direct release to the atmosphere. The extended compliance date is to allow for the availability of REC equipment.

Table A-1. (Continued)

Source Sector	Affected Facility	Prior Rules	New Rules	Compliance Dates	Comment
					Of note, the rule provides that if using REC technology is "infeasible" or "unsafe," the operator may use a completion combustion device to control flowback emissions.
			Notification and annual reporting requirements.	October 15, 2012.	Owners or operators of hydraulically fractured and refractured natural gas wells must notify EPA (or in some cases, a state or local air agency) by e-mail no later than two days before completion work begins. Further, each year, owners/ operators must submit a report on all well completion activities during the year.
	Refractured Natural Gas Well Completions	Not covered.	All wells that are refractured or recompleted prior to January 1, 2015: the well will not be considered "modified" under the Clean Air Act if operators use reduced emissions completions (i.e., "green completion") rather than flaring.	Prior to January 1, 2015.	The provisions are designed to encourage operators to use REC technology before it is required, and will allow operators in many states to refracture wells without triggering additional state permitting requirements. EPA believes that this provision will result in an additional 1,000 to 1,500 REC.

Source Sector	Affected Facility	Prior Rules	New Rules	Compliance Dates	Comment
			All wells that are refractured or recompleted on or after January 1, 2015: operators must use reduced emissions completion (i.e., "green completion").	On or after January 1, 2015.	The provisions are equivalent to new fractured wells.
	New & Modified Compressors	Not covered.	Not covered.	N/a.	N/a.
	New & Modified Pneumatic Controllers	Not covered.	Natural gas-driven, continuous-bleed devices with bleed rate greater than 6 standard cubic feet per hour (scfh): operators must reduce natural gas bleed rate to less than 6 scfh.	1-year phase-in period beginning October 15, 2012.	The provisions are designed to exclude pneumatic controllers that meet the NSPS standard (e.g., lowbleed, intermittent-bleed, and/or non-gas-driven devices), exempting them from the recordkeeping and reporting requirements of the rule to encourage operators to install.
	New Storage Vessels	Not covered.	Vessels with VOCs emissions equal to or greater than 6 tons per year (tpy): operators must reduce VOCs emissions by 95%, employing means such as a floating roof or a closed vent system and control device (e.g., flare).	1-year phase-in period beginning October 15, 2012.	Storage vessel defined as a tank or other vessel that is designed to contain an accumulation of crude oil, condensate, intermediate hydrocarbon liquids, or produced water, that is constructed primarily of nonearthen materials (such as wood, concrete, steel, fiberglass, or plastic) which provides structural support and is designed to contain an accumulation of liquids or other materials.

Table A-1. (Continued)

Source Sector	Affected Facility	Prior Rules	New Rules	Compliance Dates	Comment
					It excludes skidmounted vessels and mobile vessels attached to vehicles that are intended to remain at a site for less than 180 consecutive days, as well as process vessels, and pressure vessels designed to operate in excess of 204.9 kilopascals with no emissions to the atmosphere. The extended compliance date is to allow for the availability of control devices.
Natural Gas Gathering & Boosting Stations	New & Modified Compressors	Not covered.	Centrifugal compressors with wet-seals: operators must utilize a control system that captures emissions from the wet seal fluid degassing system and routes them to a control device in order to reduce VOCs emissions by at least 95%.	October 15, 2012.	Centrifugal compressors with dry-seals are exempt from the new rules due to low VOCs emissions.
			Reciprocating compressors: operators required to change rod packings after either 26,000 hours (operating hours must be monitored and documented) or 36 months of operation.	October 15, 2012.	Operating hours must be monitored and documented for usage by hour, but monitoring is waived if operators choose the 36 month requirement.

Source Sector	Affected Facility	Prior Rules	New Rules	Compliance Dates	Comment
	New & Modified Pneumatic Controllers	Not covered.	Natural gas-driven, continuous-bleed devices with bleed rate greater than 6 standard cubic feet per hour (scfh): operators must reduce natural gas bleed rate to less than 6 scfh.	1-year phase-in period beginning October 15, 2012.	The provisions are designed to exclude pneumatic controllers that meet the NSPS standard (e.g., lowbleed, intermittent-bleed, and/or non-gas-driven devices), exempting them from the recordkeeping and reporting requirements of the rule to encourage operators to install.
	New Storage Vessels	Not covered.	Vessels with VOCs emissions equal to or greater than 6 tons per year (tpy): operators must reduce VOCs emissions by 95%, employing means such as a floating roof or a closed vent system and control device (e.g., flare).	1-year phase-in period beginning October 15, 2012.	(See definition of "Storage Vessel" above.)
Natural Gas Processing Plants	New & Modified Compressors	Covered under LDAR program	Centrifugal compressors with wet-seals: operators must utilize a control system that captures emissions from the wet seal fluid degassing system and routes them to a control device in order to reduce VOCs emissions by at least 95%.	October 15, 2012.	Centrifugal compressors with dry-seals are exempt from the new rules due to low VOCs emissions.
			Reciprocating compressors: operators required to change rod packings after either 26,000 hours or 36 months of operation.	October 15, 2012.	Operating hours must be monitored and documented for usage by hour, but monitoring is waived if operators choose the 36 month requirement.

Table A-1. (Continued)

Source Sector	Affected Facility	Prior Rules	New Rules	Compliance Dates	Comment
	New & Modified Pneumatic Controllers	Covered under LDAR program.	Natural gas-driven, continuous-bleed devices: operators must reduce natural gas bleed rate to zero scfh.	October 15, 2012.	The standards for pneumatic controllers at natural gas processing plants reflect the emission level achievable from the use of non-natural gas-driven pneumatic controllers.
	New Storage Vessels	Covered under LDAR program.	Vessels with VOCs emissions equal to or greater than 6 tons per year (tpy): operators must reduce VOCs emissions by 95%, employing means such as a floating roof or a closed vent system and control device (e.g., flare).	1-year phase-in period beginning October 15, 2012.	(See definition of "Storage Vessel" above.)
	Leak Detection and Repair	The prior NSPS for equipment leaks of VOCs at natural gas processing plants	Existing NSPS requirements for LDAR revised to reflect the procedures and leak thresholds established by 40 C.F.R. 60, subpart VVa:	October 15, 2012.	EPA excludes compressors from the leak detection and repair (LDAR) requirements applicable to equipment at onshore natural gas processing plants because the rule already imposes control requirements on compressors.
		(40 C.F.R. Part 60, subpart KKK) requires compliance with	Subpart VVa lowers the leak definition for valves from 10,000 ppm to 500 ppm, and requires the monitoring of connectors, pumps,		

Source Sector	Affected Facility	Prior Rules	New Rules	Compliance Dates	Comment
		specific provisions of 40 C.F.R. Part 60, subpart VV, which is a LDAR program, based on the use of EPA Method 21 to identify equipment leaks.	pressure relief devices, and open-ended valves or lines.		
Natural Gas Transmission, Storage, and Distribution	New & Modified Compressors	Not covered.	Not covered.	N/a.	EPA states that it needs additional information in order to set cost-effective standards in this sector where VOCs content of gas is generally low.
	New & Modified Pneumatic Controllers	Not covered.	Not covered.	N/a.	EPA states that it needs additional information in order to set cost-effective standards in this sector where VOCs content of gas is generally low.
	New Storage Vessels	Not covered.	Vessels with VOCs emissions equal to or greater than 6 tons per year (tpy): operators must reduce VOCs emissions by 95%,	1-year phase-in period beginning October 15, 2012.	(See definition of "Storage Vessel" above.)
			employing means such as a floating roof or a closed vent system and control device (e.g., flare).		

Table A-1. (Continued)

Source Sector	Affected Facility	Prior Rules	New Rules	Compliance Dates	Comment
Oil Well Sites	Well Completions	Not covered.	Not covered.	N/a.	EPA excludes oil wells from requirements. Only "gas wells" or "natural gas wells" defined as onshore wells drilled principally for the production of natural gas are covered.
	New & Modified Compressors	Not covered.	Not covered.	N/a.	N/a.
	New & Modified Pneumatic Controllers	Not covered.	Natural gas-driven, continuous-bleed devices with bleed rate greater than 6 standard cubic feet per hour (scfh): operators must reduce natural gas bleed rate to less than 6 scfh.	1-year phase-in period beginning October 15, 2012.	The provisions are designed to exclude pneumatic controllers that meet the NSPS standard (e.g., lowbleed, intermittent-bleed, and/or non-gas-driven devices), exempting them from the recordkeeping and reporting requirements of the rule to encourage operators to install.
	New Storage Vessels	Not covered.	Vessels with VOCs emissions equal to or greater than 6 tons per year (tpy): operators must reduce VOCs emissions by 95%, employing means such as a floating roof or a closed vent system and control device (e.g., flare).	1-year phase-in period beginning October 15, 2012.	(See definition of "Storage Vessel" above.)

Source: U.S. Environmental Protection Agency, "Oil and Natural Gas Sector: New Source Performance Standards and National Emission Standards for Hazardous Air.
Pollutants Reviews, Final Rule," 77 *Federal Register* 49489, August 16, 2012; and 40 C.F.R. Part 60, Subpart KKK.

Table A-2. New Source Performance Standards for Sulfur Dioxide in Crude Oil and Natural Gas Systems

Source Sector	Affected Facility	Prior Rules	New Rules	Compliance Dates	Comments
		40 C.F.R. Part 60, Subpart LLL	40 C.F.R. Part 60, Subpart OOOO	77 FR 49489, August 16, 2012	New or Modified Facilities Defined as Beginning Construction After August 23, 2011
Natural Gas Processing Plants	New and Modified Sweetening Units	Sweetening units at onshore affected natural gas processing plants operator required to achieve an SO_2 emission reduction efficiency determined from an equation based on the sulfur feed rate and the sulfur content of the acid gas of the affected facility.	Sweetening units at onshore affected natural gas processing plants: operators must reduce SO_2 emissions for units with sulfur production rate of at least 5 long tons per day based on sulfur feed rate and sulfur content of gas.	October 15, 2012.	Sweetening unit defined as a process device that removes hydrogen sulfide and/or carbon dioxide from the sour natural gas stream.

Source: U.S. Environmental Protection Agency, "Oil and Natural Gas Sector: New Source Performance Standards and National Emission Standards for Hazardous Air Pollutants Reviews, Final Rule," 77 Federal Register 49489, August 16, 2012; and 40 C.F.R. Part 60, Subpart LLL.

Table A-3. National Emissions Standards for Hazardous Air Pollutants in Crude Oil and Natural Gas Systems

Source Sector	Affected Facility	Prior Rules 40 C.F.R. Part 63, Subparts HH (Covering the Crude Oil and Natural Gas Production Sector) and HHH (Covering the Natural Gas Transmission and Storage Sector)	New Rules 40 C.F.R. Part 63, Subparts HH (Covering the Crude Oil and Natural Gas Production Sector) and HHH (Covering the Natural Gas Transmission and Storage Sector)	Compliance Dates 77 FR 9489, August 16, 2012	Comments New or Modified Facilities Defined as Beginning Construction After August 23, 2011
Major Sources, All Sectors	Large Dehydrators	Glycol dehydration units at major sources that have annual average natural gas throughputs of 85,000 cubic meters a day or emit an average of one ton or more of benzene annually: required to connect, through a closed vent system, each process vent to an air emission control system. The control system must reduce emissions: (1) by 95% or more of HAPs, (2) to an outlet concentration of 20 parts per million by volume (ppmv) or less (for combustion devices), or (3) to a benzene emission level of one ton per year or less.	Retains existing standards.	N/a.	The final rule eliminated the exemption for periods of startup, shutdown, and malfunction.
	Small Dehydrators	Not covered.	Small glycol dehydrators that have annual average natural gas throughputs of less than 85,000 standard cubic feet per day or emit an average of less	October 15, 2012, for new units; October 15, 2015, for existing units.	The final rule eliminated the exemption for periods of startup, shutdown, and malfunction.

Source Sector	Affected Facility	Prior Rules 40 C.F.R. Part 63, Subparts HH (Covering the Crude Oil and Natural Gas Production Sector) and HHH (Covering the Natural Gas Transmission and Storage Sector)	New Rules 40 C.F.R. Part 63, Subparts HH (Covering the Crude Oil and Natural Gas Production Sector) and HHH (Covering the Natural Gas Transmission and Storage Sector)	Compliance Dates 77 FR 9489, August 16, 2012	Comments New or Modified Facilities Defined as Beginning Construction After August 23, 2011
			than one ton of benzene annually: required to comply with unit-specific limits on annual emissions of benzene, toluene, ethylene, and xylene (collectively, "BTEX").		
	Storage Vessels	Storage vessels at major sources with the potential for flash emissions: required to have either a closed vent system that is linked to a control device that either recovers or destroys 95% or more of HAPs emissions, or a combustion device (such as flares) that reduces HAPs emissions to an outlet concentration of 20 ppmv or less.	Retains existing standards	N/a.	The final rule eliminated the exemption for periods of startup, shutdown, and malfunction. EPA chose not to promulgate new standards covering non-flash potential storage tanks (e.g., crude oil and condensate storage tanks) because it determined that additional data is necessary to establish the MACT standard.
	Leak Detection and Repair	Natural gas processing plants that are major HAPs sources: control HAPs emissions from leaks from ancillary equipment and compressors. The owner or operator of such equipment is required to implement a Leak	Under the new standards, a leak triggering the rule's LDAR requirements is strengthened to include any measurement exceeding 500 ppm of any regulated HAPs.	October 15, 2012.	The final rule eliminated the exemption for periods of startup, shutdown, and malfunction.

Table A-3. (Continued)

Source Sector	Affected Facility	Prior Rules 40 C.F.R. Part 63, Subparts HH (Covering the Crude Oil and Natural Gas Production Sector) and HHH (Covering the Natural Gas Transmission and Storage Sector)	New Rules 40 C.F.R. Part 63, Subparts HH (Covering the Crude Oil and Natural Gas Production Sector) and HHH (Covering the Natural Gas Transmission and Storage Sector)	Compliance Dates 77 FR 9489, August 16, 2012	Comments New or Modified Facilities Defined as Beginning Construction After August 23, 2011
		Detection and Repair (LDAR) program and where necessary, perform equipment modifications.			
	Associated Equipment	Associated equipment for major source determinations defined as equipment associated with an oil or natural gas exploration or production well, including all equipment from the wellbore to the point of custody transfer, except glycol dehydration units and storage vessels with the potential for flash emissions.	Associated equipment for major source determinations amended to exclude emissions from all storage vessels.	October 15, 2012.	The treatment of HAPs emissions from oil and gas production has differed from the general requirements of Section 112 of the CAA. Section 112(n)(4) provides that, notwithstanding the general definition of "major source," … "emissions from any oil or gas exploration or production well (with its associated equipment) and emissions from any pipeline compressor or pump station shall not be aggregated with emissions from other similar units, whether or not such units are in a contiguous area or under common control,

Source Sector	Affected Facility	Prior Rules 40 C.F.R. Part 63, Subparts HH (Covering the Crude Oil and Natural Gas Production Sector) and HHH (Covering the Natural Gas Transmission and Storage Sector)	New Rules 40 C.F.R. Part 63, Subparts HH (Covering the Crude Oil and Natural Gas Production Sector) and HHH (Covering the Natural Gas Transmission and Storage Sector)	Compliance Dates 77 FR 9489, August 16, 2012	Comments New or Modified Facilities Defined as Beginning Construction After August 23, 2011
					to determine whether such units or stations are major sources." Thus, for facilities that are production field facilities, only HAPs emissions from glycol dehydration units and storage vessels are aggregated for a major source determination. For facilities that are not production field facilities, HAPs emissions from all HAPs emission units are aggregated for a major source determination.

Source: U.S. Environmental Protection Agency, "Oil and Natural Gas Sector: New Source Performance Standards and National Emission Standards for Hazardous Air Pollutants Reviews, Final Rule," 77 *Federal Register* 49489, August 16, 2012; and 40 C.F.R. Part 63, Subparts HH (Covering the Crude Oil and Natural Gas Production Sector) and HHH (Covering the Natural Gas Transmission and Storage Sector).

Notes: The NESHAPs in the table cover "Major Sources" of air toxics in the Crude Oil and Natural Gas Production Sector, and in the Natural Gas Transmission and Storage Sector. "Major Source" is defined as emitting 10 or more tons of a single air toxic and 25 tons or more of a combination of toxics in a year.

Table A-4. Comparison of New Federal Air Standards for Crude Oil and Natural Gas Systems to Selected State Regulations

Source	EPA's New Air Standards	Colorado	Montana	New Mexico	North Dakota	Utah	Wyominga
Well Completions	Hydraulically fractured wildcat, delineation, and low-pressure natural gas wells must route flowback emissions to completion combustion device. All other hydraulically fractured natural gas wells must route flowback emissions to completion combustion device or use green completions prior to January 1, 2015, and must use green completions after January 1, 2015, unless "infeasible."	COGCC HB-07-1341, Section 805.b(3). Green completions shall be used when technically and economically feasible. If not feasible, Best Management Practices shall be used.	MT DNRC BOGC 36.22.1221. All gas vented to the atmosphere at a rate exceeding 20 MCF per day for a period in excess of 72 hours shall be burned.	None.	None.	None.	C6 S2 O&G Permitting Guidance. Green completions are required in the JPAD area and CDA in Wyoming as of August 1, 2011.

Source	EPA's New Air Standards	Colorado	Montana	New Mexico	North Dakota	Utah	Wyominga
Processing Plants	Equipment leaks at onshore natural gas processing plants require a LDAR program.	Colorado has adopted NSPS Subpart KKK on LDAR under Reg. 7, XII.G.1 (KKK applies at gas processing plants located in ozone nonattainment areas regardless of the date of construction of the affected facility).	Montana has adopted NSPS Subpart KKK on LDAR.	New Mexico has adopted NSPS Subpart KKK on LDAR.	North Dakota has adopted NSPS Subpart KKK on LDAR.	Utah has adopted NSPS Subpart KKK on LDAR.	Wyoming has adopted NSPS Subpart KKK on LDAR.
Pneumatic Controllers	Continuous bleed natural gas-driven pneumatic controllers at natural gas processing plants must meet natural gas bleedrate of zero. Continuous bleed natural gas-driven pneumatic controllers with a bleed rate greater than 6 scfh between	Reg. 7, XVIII.C.1. No or low-bleed pneumatic devices required for all new & existing applications. (exceptions allowed) (only applies in ozone nonattainment areas).	17.8.1603(1)(a). VOCs vapors (> 500 BTU/scf) from O&G wellhead equipment must be captured and routed to a gas pipeline if within ½ mile, or to emissions minimizing technology or smokeless combustion device.	None.	None.	None.	C6 S2 O&G Permitting Guidance. Install low or no-bleed at all new facilities. Upon modification of facilities, new pneumatic controllers must be low/no-bleed and existing controllers must be replaced with no/low-bleed.

Table A-4. (Continued)

Source	EPA's New Air Standards	Colorado	Montana	New Mexico	North Dakota	Utah	Wyominga
	wellhead and natural gas processing plant must meet natural gas bleedrate less than 6 scfh.						(Well site facilities only—not gas plants.)
		COGCC HB-07-1341, Section 805.b(2)E. No or low-bleed required for new, repaired or replaced devices where technically feasible.	17.8.1711(1)(a). VOCs vapors (>200 Btu/scf) from each piece of O&G well facility equipment, with a PTE > 15 tpy, must be captured and routed to a gas pipeline, or routed to air pollution control equipment with a 95% or greater control efficiency.				
			17.8.752. Requires a case by case BACT determination.				
Compressors	Centrifugal compressors with dry seals are exempt. Centrifugal compressors with wet seals must	None.	17.8.1603(1)(a). VOCs vapors (>500 BTU/scf) from O&G wellhead equipment must be captured and routed to a gas	None.	None.	None.	None.

Source	EPA's New Air Standards	Colorado	Montana	New Mexico	North Dakota	Utah	Wyominga
	reduce emissions by 95%. Reciprocating compressors must change rod packing after 26,000 hours or after 36 months.		pipeline if within ½ mile, or to emissions minimizing technology or smokeless combustion device.				
			17.8.1711(1)(a). VOCs vapors (>200 Btu/scf) from each piece of O&G well facility equipment, with a PTE > 15 tpy, must be captured and routed to a gaspipeline, or routed to air pollution control equipment with a 95% or greater control efficiency.	None.	NDAC Section 33-15-07. Submerged filling requirements for tanks >1,000 gallons and control of organic compounds.	R307-327 Ozone Nonattainment Area. Volatile Petroleum Liquid Tanks (> 40,000 gallons, true vapor pressure (TVP) > 1.52 psia at storage temperature) shall be controlled to minimize vapor loss. New tanks shall be fitted with an internal floating roof resting on the liquid surface with the space (roof edge to tank wall) sealed.	C6 S2 O&G Permitting Guidance. JPAD - 98% control of all new/modified tank emissions upon startup/modification.

Table A-4. (Continued)

Source	EPA's New Air Standards	Colorado	Montana	New Mexico	North Dakota	Utah	Wyominga
			17.8.752. Requires a case by case BACT determination.				
Storage Vessels	Storage vessels with VOCs emissions equal to or greater than 6 tpy must reduce emissions by 95%. HAPs at production facilities must reduce emissions by 95%.	Reg. 7, XII.G.2. 95% VOCs reduction at gas processing plants if uncontrolled emissions from condensate tanks are ≥ 2 tpy (only applies in ozone non-attainment areas).	17.8.1603(1)(b). VOCs vapors from O&G oil or condensate storage tanks with a PTE > 15 tpy must be routed to a gas pipeline or emissions minimizing technology.			Owner/operator shall maintain records of the liquid type/maximum TVP. Records required of average monthly storage temperature, the liquid type, throughput and maximum TVP for tanks not subject to above (petroleum liquid TVP > 1.0 psia).	CDA - 98% control of all new/modified tank emissions ≥ 8 tpy VOCs within 60 days of startup/modification. Statewide - 98% control of all new/modified tank emissions ≥10 tpy VOCs within 60 days of startup/modification.

Source	EPA's New Air Standards	Colorado	Montana	New Mexico	North Dakota	Utah	Wyominga
		Reg. 7, XVII.C.1. 95% VOCs reduction for condensate storage tanks if uncontrolled emissions ≥ 20 tpy.	Registration - 17.8.1711 (1)(a). VOCs vapors from each piece of O&G well facility equipment with PTE >15 tpy must be captured and routed to a gas pipeline, or routed to air pollution control equipment with a 95% or greater control efficiency.				
		Reg. 7, XIID. Condensate tanks in ozone nonattainment areas shall be controlled under a system wide approach.	17.8.1711(1)(b). Requires submerged filling technology on all hydrocarbon liquid loading or unloading.				
		COGCC HB-07-1341, Section 805.b(2)A). 95% VOCs reduction for liquids condensate & crude oil tanks					

Table A-4. (Continued)

Source	EPA's New Air Standards	Colorado	Montana	New Mexico	North Dakota	Utah	Wyominga
		if uncontrolled emissions ≥ 5 tpy within 1/4 mile of an affected building (Garfield, Mesa & Rio Blanco).					
Dehydrators	95% reduction of HAPs in all large glycol dehydrators (> 3 MMCFD or > 1 tpy benzene emissions). Small dehydrator emission limits of 4.66 E-6 grams BTEX/scmppmv (new units) or 1.1 E-4 grams BTEX/scmppmv (existing units)	Reg. 7, XII.H and XVII.D. 90% reduction of VOCs where uncontrolled VOCs emissions ≥ 15 tpy.	17.8.1603(1)(a). VOCs vapors (> 500 BTU/scf) from O&G wellhead equipment must be captured and routed to a gas pipeline if within ½ mile, or to emissions minimizing technology or smokeless combustion device.				

Source	EPA's New Air Standards	Colorado	Montana	New Mexico	North Dakota	Utah	Wyominga
		COGCC HB-07- 1341, Section 805 b(2)C). 90% reduction of VOCs required where uncontrolled VOCs emissions ≥ 5 tpy within 1/4 mile of an affected building (applies only to Garfield, Mesa & Rio Blanco Counties).	17.8.1711(1)(a). VOCs vapors (>200 Btu/scf) from each piece of O&G well facility equipment, with a PTE > 15 tpy, must be captured and routed to a gas pipeline, or routed to air pollution control equipment with a 95% or greater control efficiency.	None.	TEG units with a condenser require temperature monitoring.	None.	C6 S2 O&G Permitting Guidance. JPAD - 98% control of all new/modified dehydrator VOCs/HAPs emissions at start up.
							CDA & Statewide - PAD Facilities - 98% control upon startup. SINGLE Well Facilities - 98% control within 60 days of startup for VOCs emissions ≥6 OR 98% control within 30 days of startup for VOCs emissions ≥8 tpy.

Table A-4. (Continued)

Source	EPA's New Air Standards	Colorado	Montana	New Mexico	North Dakota	Utah	Wyoming[a]
Minor Source Permitting	Exempt	Reg. 3 Part B, II.D. Minor Source permitting required for sources with thresholds that vary by pollutant and area (generally required in non-attainment areas for criteria emissions > 1-5 tpy – required statewide for criteria emissions > 5-10 tpy – thresholds depend on the pollutant).	17.8.743 Montana Air Quality Permits (MAQP). NSR permitting required for sources with > 25 tpy PTE. 17.8.1702. A registration eligible facility may register in lieu of obtaining a MAQP.	20.2.72 NMAC. Requires permits for all sources >25 tpy of a criteria pollutant. 20.2.73 NMAC. Requires Notices of Intent for all sources >10 tpy of a criteria pollutant	None. (Registration of O&G facilities required per Chapter 33-15- 20 rules in lieu of a permit.)	UAC Rule 307-401-9. NSR permitting exempted for sources with controlled emissions below de minimis levels: PTE< 5 tpy each PM10, NOx, SOx, CO, VOCs, or single HAPs < 500 lbs per year, combined HAPs < 1 tpy.	Emissions from minor sources must be approved through permitting applied through the WAQSR Chapter 6 Section 2(a)(i) O&G Permitting Guidance. For VOCs emissions ≥8 tpy from sources other than tanks, dehydrators, pneumatic controllers and pumps, water tanks, BACT is considered on case-by-case basis.

Source: Federal summaries from U.S. Environmental Protection Agency, "Oil and Natural Gas Sector: New Source Performance Standards and National Emission Standards for Hazardous Air Pollutants Reviews, Final Rule," 77 *Federal Register* 49489, August 16, 2012. State summaries from Lee Gribovicz, *Analysis of States' and EPA Oil & Gas Air Emissions Control Requirements for Selected Basins in the Western United States*, Western Regional Air Partnership, January 8, 2012, http://www.wrapair2.org/pdf/2012-01_Final%20WRAP%20OG%20Analysis%20%2801-08%29.pdf.

a. Wyoming has three area categories: (1) Jonah-Pinedale Anticline Development (JPAD), (2) Concentrated Development Area (CDA) and (3) statewide.

APPENDIX B. COMPOSITION OF FUGITIVE AND COMBUSTED NATURAL GAS EMISSIONS

Fugitive Natural Gas

Fugitive natural gas is primarily a mixture of low molecular-weight hydrocarbon compounds that are gases at surface pressure and temperature conditions. While the principal component of natural gas is methane (CH_4), it may contain smaller amounts of other hydrocarbons, such as ethane, propane, and butane, as well as trace amounts of heavier hydrocarbons. Non-hydrocarbon gases, such as carbon dioxide (CO_2), helium (He), hydrogen sulfide (H_2S), nitrogen (N_2), and water vapor (H_2O), may also be present in any proportion to the total hydrocarbon content (see Table B-1).

Table B-1. Composition of Raw Natural Gas

Component	Abbreviation	Percent of Composition
Methane (Natural Gas)	CH4	70-90
Ethane (Natural Gas Liquid)	C2H6	0-20
Propane (Natural Gas Liquid)	C3H8	0-20
Butane (Natural Gas Liquid)	C4H10	0-20
Heavier Hydrocarbons	CxHy	0-20
Carbon Dioxide	CO2	0-8
Oxygen	O2	0-0.2
Nitrogen	N2	0-5
Hydrogen Sulfide	H2S	0-5
Rare Gases	A, He, Ne, Xe	Trace

Source: The Natural Gas Supply Association, http://www.naturalgas.org/overview /background.asp.

Combusted Natural Gas

Combusted natural gas releases carbon dioxide (CO_2), carbon monoxide (CO), nitrogen oxides (NOx), and trace amounts of sulfur dioxide (SO_2), particulate matter (PM), uncombusted methane (CH_4), VOCs, and HAPs.

Combusted natural gas is generally regarded as "cleaner" than other fossil fuels with respect to various criteria pollutants and greenhouse gas (GHG) emissions (see EPA's and EIA's data comparisons in Table B-2).

Coal, fuel oil, and petroleum-based transportation fuels are composed of more complex long-chain hydrocarbon molecules, with higher carbon ratios, and may contain higher nitrogen and sulfur contents as well as increased amounts of ash and particulate matter.

By comparison, the combustion of natural gas releases minimal amounts of sulfur dioxide and nitrogen oxides, virtually no ash or particulate matter, and lower levels of carbon dioxide, carbon monoxide, and other reactive hydrocarbons.

While comparisons among the emission profiles of natural gas, coal, fuel oil, and transportation fuels are useful in certain contexts, it should be noted that these comparisons are not wholly substitutable, as end-uses of the fuels vary across sectors and pollution control mechanisms are in place for many of the fuels' combustion activities.

Table B-2. Air Pollution Emissions by Combusted Fuel Type Pounds per Billion Btu of Energy Input

Pollutant	Natural Gas	Fuel Oil	Coal
Carbon Dioxide	117,000	164,000	208,000
Carbon Monoxide	40	33	208
Nitrogen Oxides	92	448	457
Sulfur Dioxide	1	1,122	2,591
Particulates	7	84	2,744
Formaldehyde	0.750	0.220	0.221
Mercury	0.000	0.007	0.016

Source: Energy Information Administration (EIA), Office of Oil and Gas. Carbon Monoxide: derived from EIA, *Emissions of Greenhouse Gases in the United States 1997,* Table B1, p. 106. Other Pollutants: derived from Environmental Protection Agency, *Compilation of Air Pollutant Emission Factors, Vol. 1, 1998.*

Notes: No pre- or post-combustion removal of pollutants. Bituminous coal burned in a spreader stoker is compared with No. 6 fuel oil burned in an oil-fired utility boiler and natural gas burned in an uncontrolled residential gas burner. Conversion factors are: bituminous coal at 12,027 Btu per pound and 1.64% sulfur content; and No. 6 fuel oil at 6.287 million Btu per barrel and 1.03% sulfur content—derived from Energy Information Administration, *Cost and Quality of Fuels for Electric Utility Plants* (1996).

APPENDIX C. GLOSSARY OF TERMS

Table C-1. Glossary of Terms Related to Crude Oil and Natural Gas Systems As defined by the U.S. Environmental Protection Agency in the Final Rules

Terminology	Definition
Bleed rate	The rate in standard cubic feet per hour at which natural gas is continuously vented (bleeds) from a pneumatic controller.
Centrifugal compressor	Any machine for raising the pressure of a natural gas by drawing in low pressure natural gas and discharging significantly higher pressure natural gas by means of mechanical rotating vanes or impellers. Screw, sliding vane, and liquid ring compressors are not centrifugal compressors for the purposes of this subpart.
City gate	The delivery point at which natural gas is transferred from a transmission pipeline to the local gas utility.
Completion combustion device	Any ignition device, installed horizontally or vertically, used in exploration and production operations to combust otherwise vented emissions from completions.
Compressor station	Any permanent combination of one or more compressors that move natural gas at increased pressure from fields, in transmission pipelines, or into storage.
Continuous bleed	A continuous flow of pneumatic supply natural gas to the process control device (e.g., level control, temperature control, pressure control) where the supply gas pressure is modulated by the process condition, and then flows to the valve controller where the signal is compared with the process set-point to adjust gas pressure in the valve actuator.
Dehydrator	A device in which an absorbent directly contacts a natural gas stream and absorbs water in a contact tower or absorption column (absorber).
Delineation well	A well drilled in order to determine the boundary of a field or producing reservoir.
Exploratory well	A well outside known fields or the first well drilled in an oil or gas field where no other oil and gas production exists.
Flare	A thermal oxidation system using an open (without enclosure) flame. Completion combustion devices as defined in this section are not considered flares.
Flowback	The process of allowing fluids to flow from a natural gas well following a treatment, either in preparation for a subsequent phase of treatment or in preparation for cleanup and returning the well to production.

Table C-1. (Continued)

Terminology	Definition
	The flowback period begins when material introduced into the well during the treatment returns to the surface immediately following hydraulic fracturing or refracturing. The flowback period ends with either well shut in or when the well is producing continuously to the flow line or to a storage vessel for collection, whichever occurs first.
Flow line	A pipeline used to transport oil and/or gas from the well to a processing facility, a mainline pipeline, reinjection, or other useful purpose.
Hydraulic fracturing	The process of directing pressurized fluids containing any combination of water, proppant, and any added chemicals to penetrate tight formations, such as shale or coal formations, that subsequently require high rate, extended flowback to expel fracture fluids and solids during completions.
Hydraulic refracturing	Conducting a subsequent hydraulic fracturing operation at a well that has previously undergone a hydraulic fracturing operation.
Intermittent/snap-action pneumatic controller	A pneumatic controller that vents non-continuously.
Low pressure gas well	A well with reservoir pressure and vertical well depth such that 0.445 times the reservoir pressure (in psia) minus 0.038 times the vertical well depth (in feet) minus 67.578 psia is less than the flow line pressure at the sales meter.
Natural gas-driven pneumatic controller	A pneumatic controller powered by pressurized natural gas.
Natural gas liquids	The hydrocarbons, such as ethane, propane, butane, and pentane that are extracted from field gas.
Natural gas processing plant	Any processing site engaged in the extraction of natural gas liquids from field gas, fractionation of mixed natural gas liquids to natural gas products, or both.
Natural gas transmission	The pipelines used for the long distance transport of natural gas (excluding processing). Specific equipment used in natural gas transmission includes the land, mains, valves, meters, boosters, regulators, storage vessels, dehydrators, compressors, and their driving units and appurtenances, and equipment used for transporting gas from a production plant, delivery point of purchased gas, gathering system, storage area, or other wholesale source of gas to one or more distribution area(s).
Natural gas well	An onshore well drilled principally for production of natural gas.

Terminology	Definition
Non natural gas-driven pneumatic controller	An instrument that is actuated using other sources of power than pressurized natural gas; examples include solar, electric, and instrument air.
Onshore	All facilities except those that are located in the territorial seas or on the outer continental shelf.
Pneumatic controller	An automated instrument used for maintaining a process condition such as liquid level, pressure, delta-pressure and temperature.
Pressure vessel	A storage vessel that is used to store liquids or gases and is designed not to vent to the atmosphere as a result of compression of the vapor headspace in the pressure vessel during filling of the pressure vessel to its design capacity.
Process unit	Components assembled for the extraction of natural gas liquids from field gas, the fractionation of the liquids into natural gas products, or other operations associated with the processing of natural gas products. A process unit can operate independently if supplied with sufficient feed or raw materials and sufficient storage facilities for the products.
Reciprocating compressor	A piece of equipment that increases the pressure of a process gas by positive displacement, employing linear movement of the driveshaft.
Reciprocating compressor rod packing	A series of flexible rings in machined metal cups that fit around the reciprocating compressor piston rod to create a seal limiting the amount of compressed natural gas that escapes to the atmosphere.
Reduced emissions completion	A well completion following fracturing or refracturing where gas flowback that is otherwise vented is captured, cleaned, and routed to the flow line or collection system, re-injected into the well or another well, used as an on-site fuel source, or used for other useful purpose that a purchased fuel or raw material would serve, with no direct release to the atmosphere.
Salable quality gas	Natural gas that meets the composition, moisture, or other limits set by the purchaser of the natural gas, regardless of whether such gas is sold.
Storage vessel	A unit that is constructed primarily of non-earthen materials (such as wood, concrete, steel, fiberglass, or plastic) which provides structural support and is designed to contain an accumulation of liquids or other materials. The following are not considered storage vessels: (1) Vessels that are skid-mounted or permanently attached to something that is mobile (such as trucks, railcars, barges or ships), and are intended to be located at a site for less than 180 consecutive days. (2) Process vessels such as surge control vessels, bottoms receivers or knockout vessels. (3) Pressure vessels designed to operate in excess of 204.9 kilopascals and without emissions to the atmosphere.

Table C-1. (Continued)

Terminology	Definition
Sweetening unit	A process device that removes hydrogen sulfide and/or carbon dioxide from a "sour" natural gas stream.
Underground storage vessel	A storage vessel stored below ground.
Well	An oil or gas well, a hole drilled for the purpose of producing oil or gas, or a well into which fluids are injected.
Well completion	The process that allows for the flowback of petroleum or natural gas from newly drilled wells to expel drilling and reservoir fluids and tests the reservoir flow characteristics, which may vent produced hydrocarbons to the atmosphere via an open pit or tank.
Well completion operation	Any well completion with hydraulic fracturing or refracturing occurring at a gas well affected facility.
Well site	One or more areas that are directly disturbed during the drilling and subsequent operation of, or affected by, production facilities directly associated with any oil well, gas well, or injection well and its associated well pad.
Wellhead	The piping, casing, tubing and connected valves protruding above the earth's surface for an oil and/or natural gas well. The wellhead ends where the flow line connects to a wellhead valve. The wellhead does not include other equipment at the well site except for any conveyance through which gas is vented to the atmosphere.

Source: U.S. Environmental Protection Agency, "Oil and Natural Gas Sector: New Source Performance Standards and National Emission Standards for Hazardous Air Pollutants Reviews, Final Rule," 77 *Federal Register* 49489, August 16, 2012.

End Notes

[1] For more discussion on natural gas resources and technologies, see CRS Report R42814, *Natural Gas in the U.S. Economy: Opportunities for Growth*, by Robert Pirog and Michael Ratner, and CRS Report R42333, *Marcellus Shale Gas: Development Potential and Water Management Issues and Laws*, by Mary Tiemann et al. Hydraulic fracturing (hydrofracking, fracking, or fracing) is commonly defined as an oil or gas well completion process which directs pressurized fluids containing any combination of water, proppant, and any added chemicals to penetrate tight rock formations, such as shale or coal formations, in order to stimulate the oil or gas residing in the formation, and which subsequently requires high rate, extended flowback to expel fracture fluids and solids. Unconventional resources are commonly defined as follows: Tight sands gas is natural gas trapped in unusually impermeable, hard rock, or in a sandstone or limestone formation that is unusually

impermeable and non-porous (tight sand). Shale gas is natural gas trapped in shale deposits, a very fine-grained sedimentary rock, which is easily breakable into thin, parallel layers. Coal-bed methane is trapped underground in coal seams, and is generally not released into the atmosphere until coal mining activities unleash it. These resources are referred to as "unconventional" because, in the broadest sense, they are more difficult and/or less economical to extract than "conventional" natural gas, usually because the technology to reach it has not been developed fully, or is too expensive. For a more detailed discussion, see the Natural Gas Supply Association's website, http://www.naturalgas.org/overview /unconvent_ng_resource.asp.

[2] U.S. Energy Information Administration, *Annual Energy Outlook, 2012.*

[3] For example, see discussion in Nathan Hultman et al., "The Greenhouse Impact of Unconventional Gas for Electricity Generation," *Environmental Research Letters*, vol. 6 (2011), http://iopscience.iop.org/1748-9326/6/4/044008.

[4] Support for the natural gas industry has come also from the Obama Administration. In his 2012 State of the Union speech, President Obama stated, "We have a supply of natural gas that can last America nearly 100 years, and my administration will take every possible action to safely develop this energy." President Barack Obama, "Remarks by the President in State of the Union Address," Washington, DC, January 24, 2012, http://www.whitehouse.gov/the-press-office/2012/01/24/remarks-president-state-union-address.

[5] These values are averages based on carbon dioxide emitted per unit of energy generated, for more discussion on emissions by combusted fuel type, see Table B-2.

[6] For more discussion, see CRS Report R41760, *Hydraulic Fracturing and Safe Drinking Water Act Issues*, by Mary Tiemann and Adam Vann.

[7] In the case of NEPA, federal law is largely implemented by federal agencies. For more discussion, see CRS Report R42502, *Hydraulic Fracturing and the National Environmental Policy Act (NEPA): Selected Issues*, by Brandon J. Murrill.

[8] The regulations were signed by the EPA Administrator on April 17, 2012, and appeared in the Federal Register on August 16, 2012 (77 *Federal Register* 49489). For a summary of the August 16, 2012 federal air standards, see section "New Federal Air Standards for Crude Oil and Natural Gas Systems" as well as Appendix A for more detail.

[9] For more discussion, see CRS Report R40806, *Energy Projects on Federal Lands: Leasing and Authorization*, by Adam Vann.

[10] Natural gas can also be recovered as a by-product from various other sources including mining, industrial, or agricultural processes. These secondary sources are not discussed in this report. For a more detailed description of the crude oil and natural gas industry, see CRS Report R40872, *U.S. Fossil Fuel Resources: Terminology, Reporting, and Summary*, by Carl E. Behrens, Michael Ratner, and Carol Glover.

[11] Crude oil after the production phase (i.e., beginning with petroleum refining) is classified as a different industrial sector for most regulatory purposes. Petroleum refining is not discussed in this report. See CRS Report R41478, *The U.S. Oil Refining Industry: Background in Changing Markets and Fuel Policies*, by Anthony Andrews, Robert Pirog, and Molly F. Sherlock.

[12] The description of the natural gas sector is drawn from the U.S. Environmental Protection Agency, "Oil and Natural Gas Sector: New Source Performance Standards and National Emission Standards for Hazardous Air Pollutants Reviews," 76 *Federal Register* 52738, August 23, 2011.

[13] For further discussion on emissions estimates, see section "Measurement of Emissions."

[14] EPA categorizes emissions as either "equipment leaks and vented emissions" or "combustion-related emissions." See U.S. Environmental Protection Agency, *Inventory of U.S. Greenhouse Gas Emissions and Sinks: 1990-2009*, Washington, DC, EPA 430-R-11-005, April 2011.

[15] For definitions of fugitive emissions, see U.S. Environmental Protection Agency, *Inventory of U.S. Greenhouse Gas Emissions and Sinks: 1990-2009*, op cit.; and Intergovernmental Panel on Climate Change, *2006 IPCC Guidelines for National Greenhouse Gas Inventories*, prepared by the National Greenhouse Gas Inventories Programme, Simon Eggleston, et al. (eds), Institute for Global Environmental Strategies, Japan, 2006, http://www.ipcc-nggip.iges.or.jp/ public/2006gl/pdf/0_Overview/V0_0_Cover.pdf.

[16] For more discussion on the chemical composition of raw natural gas emissions, see Appendix B.

[17] For more discussion on emissions from combusted fossil fuel resources, see Appendix B. While there is general agreement that the combustion of natural gas produces less GHG emissions than other combusted fossil fuels, the GHG emission intensity for the full fuel life-cycle of natural gas (i.e., from extraction through combustion), compared to petroleum and coal is under greater debate. For further discussion on the comparable impacts of the natural gas industry on GHG emissions, see section "Greenhouse Gases."

[18] 42 U.S.C. 7401 *et seq*.

[19] See authorities as defined in 42 U.S.C. 7409.

[20] See authorities as defined in 42 U.S.C. 7402.

[21] See authorities as defined in U.S. Environmental Protection Agency, "Endangerment and Cause or Contribute Findings for Greenhouse Gases," 74 *Federal Register* 66496-66516, December 15, 2009.

[22] The "endangerment" language in §§108, 111, 211, 213, 115, and 231 provides fundamental authorities. Also, Section 111(d) provides authority to control GHG emissions from existing sources and Section 111(b) and (e) provide similar authorities for new sources.

[23] Ozone is not an air pollutant that is directly emitted, but rather is formed in the atmosphere by the interaction of VOCs and NO_x in the presence of sunlight. The control of ozone is, thus, based on regulating emissions of VOCs and NO_x.

[24] The 2008 National Emissions Inventory estimated the crude oil and natural gas production sector's VOCs emissions at 1.7 million tons. Mobile sources are the highest category for VOCs emissions domestically, at 45.2% in 2008. Data for VOCs, as well as the other criteria and HAPs, are derived from EPA's National Emissions Inventory, and can be found at http://www.epa.gov/ttn/chief/eiinformation.html.

[25] U.S. Environmental Protection Agency, *Regulatory Impact Analysis: Final National Ambient Air Quality Standards for Ozone*, Research Triangle Park, NC, July 2011.

[26] NO_x emissions from engines and turbines are covered by the Standards of Performance for Stationary Spark Internal Combustion Engines (40 C.F.R. §60, subpart JJJJ) and Standards of Performance for Stationary Combustion Turbines (40 C.F.R. §60, subpart KKKK), respectively.

[27] U.S. Environmental Protection Agency, *Particulate Matter Health* website, http://www.epa.gov/pm/health.html. For more discussion, see CRS Report R40096, *2006 National Ambient Air Quality Standards (NAAQS) for Fine Particulate Matter (PM2.5): Designating Nonattainment Areas*, by Robert Esworthy.

[28] Hydrogen sulfide was on the original list of hazardous air pollutants in the CAA but was subsequently removed by Congress. See U.S. Government Accountability Office, *Clean Air Act: EPA Should Improve the Management of Its Air Toxics Program*, GAO-06-669,

Washington, DC, June 2006. According to EPA, there are 14 major areas found in 20 different states where hydrogen sulfide is commonly found in natural gas deposits. As a result of drilling in these areas, "the potential for routine [hydrogen sulfide] emissions is significant." See U.S. Environmental Protection Agency, *Report to Congress on Hydrogen Sulfide Air Emissions Associated with the Extraction of Oil and Natural Gas,* EPA-453/R-93-045, Research Triangle Park, NC, October 1993, at ii, III-35; see also ii, II-5 to II-11.

[29] U.S. Environmental Protection Agency, *Air Quality Criteria for Ozone and Related Photochemical Oxidants (2006 Final),* Washington, DC, EPA-600/R-05/004aF-cF, 2006. As an ozone precursor, methane emissions are different than NO_x and VOCs, which affect ozone concentrations regionally and at hourly time scales. Methane reductions require several decades for the ozone response to be fully realized, given methane's relatively long atmospheric lifetime (see Task Force on Hemispheric Transport of Air Pollution (HTAP). *Hemispheric Transport of Air Pollution 2010.* Informal Document No.10. Convention on Long-range Transboundary Air Pollution Executive Body 28th Session. ECE/EB.AIR/2010/10 (Corrected). Chapter 4, pp. 148-149.) The health benefits of ozone reduction are described in preceding sections.

[30] As a GHG, methane emitted into the atmosphere absorbs terrestrial infrared radiation which contributes to increased global warming and continuing climate change. Methane's role as a ground-level ozone precursor also impacts global temperatures. According to the Intergovernmental Panel on Climate Change (IPCC) *Fourth Assessment Report (2007),* in 2004 the cumulative changes in methane concentrations since preindustrial times contributed about 14% to global warming, making methane the second leading long-lived climate forcer after CO2 globally. The perturbation lifetime for methane is 12 years. CO2, however, does not undergo a simple decline over a single predictable timescale. Instead, the excess carbon is first diluted by the carbon cycle as it mixes into the oceans and biosphere (e.g., plants) over a period of a few hundred years, and then it is slowly removed over hundreds of thousands of years as it is gradually incorporated into carbonate rocks.

[31] U.S. Environmental Protection Agency, *Inventory of U.S. Greenhouse Gas Emissions and Sinks: 1990-2010,* Washington, DC, EPA 430-R-12-001, April 15, 2012, http://www.epa.gov/climatechange/Downloads/ghgemissions/ US-GHG-Inventory-2012-Main-Text.pdf. Greenhouse gases include carbon dioxide (CO2), methane (CH4), nitrous oxide (N2O), hydrofluorocarbons (HFCs), perfluorocarbons (PFCs), and sulfur hexafluoride (SF6), among many others. GHG are quantified using a unit measurement called carbon dioxide equivalent, wherein the various gases are indexed and aggregated against one unit of CO2. Thus, a CO2-equivalent quantity would have the same effect on, say, radiative forcing of the climate, as the same quantity of CO2.

[32] U.S. Environmental Protection Agency, *Regulatory Impact Analysis: Final New Source Performance Standards and Amendments to the National Emissions Standards for Hazardous Air Pollutants for the Oil and Natural Gas Industry,* Research Triangle Park, NC, April 2012, http://www.epa.gov/ttn/ecas/regdata/RIAs/ oil_natural_gas_final_neshap_nsps_ria.pdf. Agency, industry, and academic literature on the subject of estimates and impacts of emissions is substantial, but is not in agreement. For further discussion, see sections "Measurement of Emissions" and "Impacts of Emissions" of this report.

[33] For further discussion on climate change and its potential impacts, see CRS Report RL34266, *Climate Change: Science Highlights*, by Jane A. Leggett.

[34] For more discussion on the process of hydraulic fracturing, see CRS Report R42333, *Marcellus Shale Gas: Development Potential and Water Management Issues and Laws*, by Mary Tiemann et al.

[35] Specifically, EPA estimates uncontrolled well completion emissions for a hydraulically fractured well at about 23 tons of VOCs, and emissions for a conventional gas well completion at around 0.1 ton VOCs. See EPA, "Oil and Natural Gas Sector: New Source Performance Standards and National Emission Standards for Hazardous Air Pollutants Reviews," 76 *Federal Register* 52738, August 23, 2011.

[36] Reduced emission completions (REC), which are sometimes referred to as "green completions" or "flareless completions," use equipment at the well site to capture and treat gas so it can be directed into the sales line and avoid emissions from venting. Based on information provided to EPA Natural Gas STAR program, 90% of gas potentially vented during a completion can be recovered during a reduced emission completion. For more discussion of REC and a schematic, see Figure 3.

[37] For further discussion on emissions quantities during flowback, see section "Measurement of Emissions."

[38] 42 U.S.C. 7401 *et seq*. For a summary of the Clean Air Act and EPA's air and radiation activities and its authorities, see EPA's website at http://www.epa.gov/air/basic.html; or CRS Report RL30853, *Clean Air Act: A Summary of the Act and Its Major Requirements*, by James E. McCarthy et al.

[39] "Outdoor" defined as that to which the public has access (see 40 C.F.R. §50.1(e)).

[40] 42 U.S.C. §§7661-7661f; 40 C.F.R. §§71, 72.

[41] EPA's guidance for "major source" determinations includes consideration of proximity, ownership, and industrial grouping. For further discussion, see section "Major Source Aggregation."

[42] U.S. Environmental Protection Agency, "Mandatory Reporting of Greenhouse Gases," 74 *Federal Register* 56260, October 30, 2009.

[43] U.S. Environmental Protection Agency, "Mandatory Reporting of Greenhouse Gases: Petroleum and Natural Gas Systems," 75 *Federal Register* 74458, November 30, 2010, see final rule revision to Subpart W—Petroleum and Natural Gas Systems—amending 40 C.F.R. §98 (i.e., the regulatory requirements for the Program).

[44] In the final rule (40 C.F.R. §98.238), EPA defined "facility" with respect to onshore petroleum and natural gas production to mean "all petroleum or natural gas equipment on a well pad or associated with a well pad and CO2 EOR [enhanced oil recovery] operations that are under common ownership or common control including leased, rented, or contracted activities by an onshore petroleum and natural gas production owner or operator and that are located in a single hydrocarbon basin [i.e., as defined on a county level and by geologic formation]. Where a person or entity owns or operates more than one well in a basin, then all onshore petroleum and natural gas production equipment associated with all wells that the person or entity owns or operates in the basin would be considered one facility."

[45] *WildEarth Guardians, et al. v. Jackson,* No. 1:09-CV-00089-CKK (D. D.C.), https://www.federalregister.gov/ articles/2009/12/17/E9-30044/proposed-consent-decree-clean-air-act-citizen-suit.

[46] U.S. Environmental Protection Agency, "Oil and Natural Gas Sector: New Source Performance Standards and National Emission Standards for Hazardous Air Pollutants Reviews," 76 *Federal Register* 52738, August 23, 2011.

[47] U.S. Environmental Protection Agency, "Oil and Natural Gas Sector: New Source Performance Standards and National Emission Standards for Hazardous Air Pollutants Reviews, Final Rule," 77 *Federal Register* 49489, August 16, 2012, https://www.federalregister.gov/articles/2012/08/16/2012-16806/oil-and-natural-gas-sector-new-source-performance-standards-and-national-emission-standards-for.

[48] Exploratory well is defined as a well outside known fields or the first well drilled in an oil or gas field where no other oil and gas production exists; delineation well is defined as a well drilled in order to determine the boundary of a field or producing reservoir; and low pressure well is defined as a well with reservoir pressure and vertical well depth such that 0.445 times the reservoir pressure (in psia) minus 0.038 times the vertical well depth (in feet) minus 67.578 psia is less than the flow line pressure at the sales meter.

[49] The final MACT standards for small dehydrators at crude oil and natural gas production facilities require that existing affected sources at a major source meet a unit-specific BTEX limit of 3.28×10^{-4} grams BTEX/standard cubic meters (scm)-parts per million by volume (ppmv) and that new affected sources meet a BTEX limit of 4.66×10^{-6} grams BTEX/scm-ppmv.

[50] As reported in U.S. Environmental Protection Agency, "Oil and Natural Gas Sector: New Source Performance Standards and National Emission Standards for Hazardous Air Pollutants Reviews, Final Rule," 77 *Federal Register* 49489, August 16, 2012.

[51] Energy Information Administration, *U.S. Natural Gas Wellhead Value and Marketed Production,* http://www.eia.gov/dnav/ng/ng_prod_whv_dcu_nus_a.htm.

[52] For further discussion on costs, see section "Cost Benefit Analysis of Federal Standards."

[53] For an example of industry comments on the cost analysis, see comments to "Oil and Natural Gas Sector: New Source Performance Standards and National Emission Standards for Hazardous Air Pollutants Reviews," submitted by Howard J. Feldman, Director, Regulatory and Scientific Affairs, American Petroleum Institute, November 30, 2011, accessed on http://www.regulations.gov.

[54] For an example of environmental stakeholder comments on the cost analysis, see comments to "Oil and Natural Gas Sector: New Source Performance Standards and National Emission Standards for Hazardous Air Pollutants Reviews," submitted by Sierra Club, et al., November 30, 2011, accessed on http://www.regulations.gov.

[55] 77 *Federal Register* 49489, *op cit.*

[56] For example, see New Mexico's current efforts to loosen regulations on pits for waste or drill cuttings and the use of steel tanks as part of a "closed loop" system, Mike Soraghan, "N.M. is loosening drilling rules, bucking trends and riling ranchers," *E&E News,* Thursday, November 15, 2012, http://www.eenews.net/energywire/2012/11/15/1.

[57] For example, see the comments made by Wyoming governor Matthew Mead, as reported in *The Washington Times:* Matthew Mead, "Commentary: Hydro-fracking regulations should be left to states," *The Washington Times*, September 17, 2012, http://www.washingtontimes.com/news/2012/sep/17/hydro-fracking-regulations-should-be-left-to-state/ #ixzz26uwFzyr3.

[58] For example, see the comments made by Peter Zalzal, attorney for the Environmental Defense Fund, "National Clean Air Standards For The Oil And Gas Industry Provide a Trifecta," EDF Energy Exchange, March 20, 2012, http://blogs.edf.org/energyexchange /2012/03/20/national-clean-air-standards-for-the-oil-and-gas-industry-provide-atrifecta/.

[59] As reported in Mike Soraghan, "Puny fines, scant enforcement leave drilling violators with little to fear," *E&E News,* Monday, November 14, 2011, http://www.eenews.net/Greenwire /2011/11/14/archive/1; Mike Soraghan, "40% of state drilling regulators have industry ties," *E&E News,* Monday, December 19, 2011, http://www.eenews.net/Greenwire/ground_rules/2011/12/19/1. The reporting found that in Texas, 96% of the 80,000 violations by oil and gas drillers in 2009 resulted in no enforcement action. West Virginia, a state with 56,000 wells, issued 19 penalties last year. And Wyoming, the center of Rocky Mountain development, collected $15,500 in fines in 2010. Pennsylvania sought penalties for more than a quarter of the violations found last year, but levied fines for only 4% of the

violations, with the penalties totaling $3.7 million. Further, the reporting reviewed the backgrounds of 95 oil and gas commissioners, board members and agency heads in the top 27 oil and gas states, and found, 39 had an oil and gas background, or 41%.

[60] Lisa Sumi, *Breaking All the Rules: The Crisis in Oil & Gas Regulatory Enforcement*, Earthworks, Washington, DC, September 25, 2012, http://www.earthworksaction.org /library/detail/ breaking_all_the_rules_the_crisis_in_oil_andjas_regulatory_enforcement.

[61] Reporting would take place under the EPA's Greenhouse Gas Reporting Program. See U.S. Environmental Protection Agency, "Mandatory Reporting of Greenhouse Gases: Petroleum and Natural Gas Systems," 75 *Federal Register* 74458, November 30, 2010, see final rule revision to Subpart W—Petroleum and Natural Gas Systems— amending 40 C.F.R. §98 (i.e., the regulatory framework for the Greenhouse Gas Reporting Program (GHGRP)).

[62] See Texas Commission on Environmental Quality, *Barnett Shale Special Inventory,* http://www.tceq.texas.gov/ assets/public/implementation/air/ie/pseiforms/summarydatainfo.pdf.

[63] See Pennsylvania Department of Environmental Protection announced protocol, http://files.dep.state.pa.us/Air/ AirQuality/AQPortalFiles/Long-Term_Marcellus_Ambient_ Air_Monitoring_Project-Protocol_for_Web_2012-07- 23.pdf.

[64] See California Environmental Protection Agency, Air Resources Board, *2007 Oil and Gas Industry Survey Results, Final Report,* December 2011, and other relevant documents on the ARB website, http://www.arb.ca.gov/cc/oil-gas/oilgas.htm.

[65] The University of Texas at Austin, along with two environmental engineering firms, URS and Aerodyne Research, announced on October 10, 2012, the launch of a comprehensive study of methane emissions around gas wells. Researchers are to focus on gas production sites in the Marcellus, Eagle Ford, Haynesville, Barnett, Fayetteville, and Niobrara shale formations. Nine oil and gas companies have agreed to allow sampling teams onto their sites: Anadarko Petroleum Corp., BG Group PLC, Chevron Corp., Encana Oil & Gas (USA) Inc., Pioneer Natural Resources Co., Shell Oil Co., Southwestern Energy, Talisman Energy and XTO Energy, a subsidiary of Exxon Mobil Corp. The announced study is part of a broader five-part investigation to measure methane leakage across the natural gas supply chain, http://www.utexas.edu/news/2012/10/10/university-of-texas-at-austin-study-measures-methane-emissions-releasedfrom-natural-gas-production/.

[66] Secretary of Energy Advisory Board Shale Gas Production Subcommittee, *Shale Gas Production Subcommittee 90 Day Report*, p.16.

[67] Gas Research Institute and U.S. Environmental Protection Agency, *Methane Emissions from the Natural Gas Industry, Volumes 1-15*, GRI-94/0257 and EPA 600/R-96-080, June 1996.

[68] UNFCCC, Treaty Number: 102-38, October 7, 1992, the resolution of advice and consent to ratification agreed to in the Senate by Division Vote. For more discussion of U.S. commitments, see CRS Report R40001, *A U.S.-Centric Chronology of the International Climate Change Negotiations*, by Jane A. Leggett.

[69] 314 bscf of methane is equivalent to approximately 127 million metric tons of carbon dioxide equivalent (MMtCO2e) greenhouse gases, using the conversion 1000 bscf CH4 = 0.4045 MMtCO2e at 60 degrees Fahrenheit (15.6 degrees Celsius) and either 14.696 psi (1 atm or 101.325 kPa) or 14.73 psi (30 inHg or 101.6 kPa) of pressure.

[70] See U.S. Environmental Protection Agency, *Inventory of U.S. Greenhouse Gas Emissions and Sinks: 1990-2010*, Washington, DC, EPA 430-R-12-001, April 15, 2012, http://www.epa.gov/climatechange/Downloads/ghgemissions/ US-GHG-Inventory-2012-Main-Text.pdf. Conversion factor of 1000 bscf CH4 = 0.4045 MMtCO2e at 60 degrees Fahrenheit (15.6 degrees Celsius) and either 14.696 psi (1 atm or 101.325 kPa) or 14.73 psi (30 inHg or

101.6 kPa) of pressure. The U.S. Energy Information Administration reports 2010 U.S. natural gas marketed production as 22,402 bscf, see http://www.eia.gov /dnav/ng/hist/n9050us2a.htm.

[71] The figures were drawn from four studies over two workshops that included data from over 1,000 wells completed between 2002 and 2006. U.S. Environmental Protection Agency, "Green Completions," Natural Gas STAR Producers' Technology Transfer Workshop, September 21, 2004 and U.S. EPA, "Reducing Methane Emissions During Completion Operations," Natural Gas STAR Producers' Technology Transfer Workshop, September 11, 2007. The presentations included a Devon case study from the Fort Worth Basin (30 wells, 11,900 Mcf/completion), a Williams case study from the Piceance Basin (1,064 wells, 24,449 Mcf/completion), a Weatherford coal bed methane case study (3 wells, 667 Mcf/completion), and a nationwide industry data set (106 wells, 5,820 Mcf/completion). For more on the EPA Natural Gas STAR program, see http://www.epa.gov/gasstar/.

[72] U.S. Environmental Protection Agency, *Greenhouse Gas Emissions Reporting from the Petroleum and Natural Gas Industry: Background Technical Support Document,* Climate Change Division, Washington, DC, 2010, http://www.epa.gov/climatechange /emissions/downloads10/Subpart-W_TSD.pdf.

[73] IHS CERA estimates methane emissions from the oil and gas "production" sector to be 43 MMtCO2e as opposed to EPA's estimate of 130 MMtCO2e from "field production." See Mary Barcella, Samantha Gross, and Surya Rajan, *Mismeasuring Methane: Estimating Greenhouse Gas Emissions from Upstream Natural Gas Development*, IHS CERA, Private Report, Cambridge, MA, 2011, http://press.ihs.com/press-release/recent-estimates-greenhouse-gas-emissionsshale-gas-production-are-likely-significantl.

[74] "Liquids unloading" is a maintenance activity in depleted conventional reservoirs where water produced with the gas has accumulated to the point where it stops gas flow. In order to reestablish gas flow, the well is blown to the atmosphere, causing the vented release of emissions. The 2011 GHG Inventory has this category listed for contributing approximately 34% of all methane emissions from the natural gas industry for that year.

[75] Terri Shires and Miriam Lev-On, *Characterizing Pivotal Sources of Methane Emissions from Unconventional Natural Gas Production: Summary and Analysis of API and ANGA Survey Responses,* Final Report, June 1, 2012, http://www.anga.us/media/249160/anga%20api%20survey%20report%201%20june%20final.pdf. The authors present data reported by over 20 of their industry partners from over 91,000 wells (or 20% of known wells in operation) demonstrating that a re-calculation of two of EPA's source categories—liquids unloading and unconventional gas re-fracture rates—returned a 50% reduction in industry emissions compared to EPA's estimate. The study did not report any new estimates for the quantity of emissions from hydraulically fractured well completions or re-completions, only estimates on the rates of re-fracture.

[76] URS, "Gas Well Completion Data," Attachment 3 of the American Exploration and Production Council and America's Natural Gas Alliance, *Comments – Proposed Rule – Oil and Natural Gas Sector Consolidated Rulemaking, NSPS and NESHAP Reviews*, November 30, 2011, http://epa.gov/quality/informationguidelines/documents/12003- attB.pdf.

[77] Darren Smith, Devon Energy Corporation, Testimony before the Senate Environment and Public Works Subcommittee on Clean Air and Nuclear Safety, Washington, DC, June 19, 2012, http://epw.senate.gov/public/ index.cfm?FuseAction=Files.View&FileStore_id= 5a14d73e-36d6-4f5b-8985-98a506e3cbab.

[78] Data reviewed by EPA showed methane concentrations ranging from 65.7% to 97.2% and VOCs concentrations from 1.2% to 5.7%. EPA estimated that escaping natural gas contains

an average of 83% methane (plus or minus 1.4% at the 95% confidence level) and 3.7% volatile organic compounds by volume (plus or minus 0.33%). From "Composition of Natural Gas for use in the Oil and Natural Gas Sector Rulemaking," memo from Heather Brown, P.E., EC/R Inc., to Bruce Moore, EPA, July 28, 2011, Table 7.

[79] See U.S. Environmental Protection Agency, *Regulatory Impact Analysis: Proposed New Source Performance Standards and Amendments to the National Emissions Standards for Hazardous Air Pollutants for the Oil and Natural Gas Industry,* Research Triangle Park, NC, July 2011, p. 4-1.

[80] Scientists typically use two different methods when trying to understand how a given air pollution source affects local air quality. One approach is to conduct ambient air monitoring, which directly measures air pollution levels that people breathe. However, in using this method, the presence of a pollutant likely reflects contributions from many different sources, such as motor vehicles, gasoline stations, and industrial plants. Another approach is to use dispersion modeling, which estimates air pollution levels using models that predict how pollutants move through the air from the point where they are released. However, in using this method, measurements of the initial outputs must be completed and verified in order to make accurate dispersion assessments. Satellite modeling and direct component measurements are other ways of quantifying emissions. These top-down and bottom-up approaches to assessments also have their specific drawbacks, including source attribution and data availability.

[81] Comments by Dr. Christopher Portier, head of CDC's National Center for Environmental Health and Agency for Toxic Substances and Disease Registry (ATSDR), as covered on January 5, 2012 by Kevin Begos of the Associated Press quoting excerpts of an email. Publication of the full text of the email, as released by the CDC press office, can be found at http://www.slopefarms.com/2012/01/08/shale-gas-drilling-and-public-health-first-publication-of-full-text-email-on-public-health-risks-of-from-cdcs-national-center-for-environmental-health-and-agency-for-toxic-substances-and-dis/.

[82] U.S. Government Accountability Office, *Oil and Gas: Information on Shale Resources, Development, and Environmental and Public Health Risks*, GAO-12-732, September 2012, pp. 33-37, http://www.gao.gov/assets/650/ 647791.pdf.

[83] Secretary of Energy Advisory Board Shale Gas Production Subcommittee, *Shale Gas Production Subcommittee 90- Day Report*, p.25. For example, as a part of New York state's ongoing review of hydraulic fracturing, the New York State Department of Environmental Conservation (NY DEC) has asked New York State Department of Health (NY DOH) to undertake a review of the NY DEC health impact analysis, see http://www.dec.ny.gov/press /85071.html.

[84] Russell and Pollack, *Oil and Gas Emission Inventories for the Western States,* Environ International Corporation study prepared for the Western Governors' Association, December 27, 2005, http://www.wrapair.org/forums/ssjf/ documents/eictts/OilGas/WRAP_Oil&Gas_Final_Report.122805.pdf.

[85] See EPA Clearinghouse for Inventories and Emissions Factors, *Compilation of Air Pollutant Emission Factors AP-42, Fifth Edition, Volume I: Stationary Point and Area Sources,* January 1995, http://www.epa.gov/ttn/chief/ap42/ch05/ final/c05s02.pdf.

[86] Colorado Dept. of Public Health & Environment, Air Pollution Control Division, *Oil and Gas Emission Sources Presentation for the Air Quality Control Commission Retreat,* May 15, 2008, at pages 3-4.

[87] See Letter from Wyoming Governor Dave Freudenthal to Carol Rushin, Acting Regional Administrator, USEPA Region 8, (Mar. 12, 2009) ("Wyoming 8Hour Ozone Designation Recommendations"), http://deq.state.wy.us/out/ downloads/Rushin%20Ozone.pdf.

[88] Wyoming Department of Environmental Quality, *Technical Support Document I for Recommended 8-hour Ozone Designation of the Upper Green River Basin,* March 26, 2009 ("Wyoming Nonattainment Analysis"), http://deq.state.wy.us/out/downloads/Ozone%20 TSD_Outside%20UGRB%20%282%29.pdf.

[89] Randal Martin, et al., *Final Report: Uinta Basin Winter Ozone and Air Quality Study, December 2010-March 2011,* Energy Dynamics Laboratory, Utah State University, for Uintah Impact Mitigation Special Service District, June 14, 2011, http://rd.usu.edu/ files/uploads/edl_2010-11_report_ozone_final.pdf.

[90] BLM, *GASCO Energy Inc. Uinta Basin Natural Gas Development Draft Environmental Impact Statement* ("GASCO DEIS"), p. 3-13, http://www.blm.gov/ut/st/en/fo /vernal/planning/nepa_/gasco_energy_eis.html.

[91] Marco Rodriguez, et al., "Regional Impacts of Oil and Gas Development on Ozone Formation in the Western United States," *Journal of the Air and Waste Management Association,* 111 (Sept. 2009), http://www.wrapair.org/forums/amc/ meetings/091111_Nox/Rodriguez_et_al_ OandG_Impacts_JAWMA9_09.pdf.

[92] See comments made by Keith Guille, spokesman for Wyoming DEQ, in Charles W. Schmidt, "Blind Rush?: Shale Gas Boom Proceeds amid Human Health Questions," *Environmental Health Perspectives*, vol. 119, no. 8 (August 2011), p. A352.

[93] See *2012 Uintah Basin Winter Ozone & Air Quality Study-Summary of Interim Findings, Ongoing Analyses, and Additional Recommended Research,* August 7, 2012, at http:// op.bna.com/env.nsf/id/smiy-8x9rv3/ $File/UintahWinter.pdf.

[94] Texas Railroad Commission, http://www.rrc.state.tx.us/data/fielddata/barnettshale.pdf.

[95] Al Armendariz, *Emissions from Natural Gas Production in the Barnett Shale Area and Opportunities for Cost-Effective Improvements,* January 26, 2009, http://www.edf.org /documents/9235_Barnett_Shale_Report.pdf. It should be noted that Al Armendariz is the former EPA Region 6 Administrator who resigned on April 30, 2012, after remarks made in 2010 regarding EPA's enforcement policy.

[96] Eduardo P. Olaguer, "The Potential Near-source Ozone Impacts of Upstream Oil and Gas Industry Emissions," *Journal of the Air & Waste Management Association*, vol. 62, no. 8 (2012), http://www.tandfonline.com/doi/abs/ 10.1080/10962247.2012.688923.

[97] Texas Commission on Environmental Quality, "A Commitment to Air Quality in the Barnett Shale," *Natural Outlook Newsletter*, Fall 2010, http://www.tceq.state.tx.us/publications/pd /020/10-04/a-commitment-to air quality-inthe-barnett-shale.

[98] Theo Colborn, Carol Kwiatkowski, Kim Schultz, Mary Bachran, "Natural Gas Operations from a Public Health Perspective," *Human and Ecological Risk Assessment*, vol. 17, no. 5, 2011, pp. 1039-1056, http://www.endocrinedisruption.com/chemicals.journalarticle.php.

[99] Amy Mall et al., *Drilling Down: Protecting Western Communities from the Health and Environmental Effects of Oil and Gas Production,* NRDC, October 2007, (citing CDPHE data), http://www.nrdc.org/land/use/down/down.pdf.

[100] Eastern Research Group, *City of Fort Worth Natural Gas Air Quality Study,* 2011, at http:// fortworthtexas.gov/ gaswells/?id=87074.

[101] Comments by Dr. Christopher Portier, CDC, *op cit.*

[102] ATSDR, *Health Consultation: Public Health Implications of Ambient Air Exposures to Volatile Organic Compounds as Measured in Rural, Urban, and Oil & Gas Development*

Areas Garfield County, Colorado, 2008, http://www.cdphe.state.co.us/dc/ehs/Garfield County.pdf.

[103] PEHSU, *Information on Natural Gas Extraction and Hydraulic Fracturing for Health Professionals*, 2011, http://aoec.org/pehsu/documents/hydraulic_fracturing_and_children_2011_health_prof.pdf.

[104] Roxana Witter, et al., *Draft Health Impact Assessment for Battlement Mesa, Garfield County, Colorado,* Colorado School of Public Health, 2011, http://www.garfield-county.com/index.aspx?page=1408; Lisa McKenzie, et al., *Human Health Risk Assessment of Air Emissions from Development of Unconventional Natural Gas Resources,* Colorado School of Public Health, 2012, http://www.energyindepth.org/wp-content/uploads/2012/03/McKenzie-CSPH-Study02-10-2012.pdf.

[105] Wolf Eagle Environmental, *Town of DISH, Texas Ambient Air Monitoring Analysis Final Report,* 2009, http://townofdish.com/objects/DISH_-_final_report_revised.pdf.

[106] Eastern Research Group, *City of Fort Worth Natural Gas Air Quality Study,* 2011, at http://fortworthtexas.gov/ gaswells/?id=87074.

[107] Barbara Zielinska, Eric Fujita, Dave Campbell, *Monitoring of Emissions from Barnett Shale Natural Gas Production Facilities for Population Exposure Assessment. Houston TX,* Desert Research Institute, 2011, http://www.sph.uth.tmc.edu/mleland/attachments/Barnett%20Shale%20Study%20Final%20Report.pdf.

[108] McKenzie, et al., *op cit.*

[109] Comments critical of the CSPH reports can be found at the Energy In Depth blog, http://www.energyindepth.org/ non-elite-eight-worst-inputs-used-in-new-colorado-health-study/.

[110] Cardno Entrix, *Hydraulic Fracturing Study: PXP Inglewood Oil Field*, Prepared for Plains Exploration & Production Company, Los Angeles, CA, October 10, 2012, http://www.inglewoodoilfield.com/res/docs/102012study/ Hydraulic%20Fracturing%20Study%20Inglewood%20Field10102012.pdf.

[111] Charles G. Groat and Thomas W. Grimshaw, *Fact-Based Regulation for Environmental Protection in Shale Gas Development*, The Energy Institute, The University of Texas at Austin, Austin, TX, February 2012, p. 30, http://energy.utexas.edu/images/ei_shale_gas_regulation120215.pdf. It should be noted that in July 2012, the objectivity of The Energy Institute study was called into question by the Public Accountability Initiative, and officials at the University of Texas have since announced their plans to commission an outside review of the study.

[112] For an example of industry responses, see the comments to "Oil and Natural Gas Sector: New Source Performance Standards and National Emission Standards for Hazardous Air Pollutants Reviews," submitted by Howard J. Feldman, Director, Regulatory and Scientific Affairs, American Petroleum Institute, November 30, 2011, accessed on http://www.regulations.gov. API asserts that establishing new and more stringent standards would be "patently unreasonable and cannot be justified ... because the so-called regulatory 'gaps' in the current rules clearly are not contributing to unacceptable risk," and that EPA's new NESHAPs are "regulation for the sake of regulation and, as such, contradict Congress's clear intent that an ample margin of safety is an appropriate stopping point for emissions limitations under § 112" of the CAA. Industry's (and others') comments were incorporated into the final rules, and the revisions are reflected and discussed in EPA's "Summary of Key Changes to the Rules" Fact Sheet, at http://www.epa.gov/airquality/oilandgas/pdfs/20120417changes.pdf.

[113] For a brief explanation of life-cycle emissions assessment methodology, see text box "Estimating the GHG Footprint of Unconventional Natural Gas."

[114] Of the 215.4 MMtCO2e of methane reported, 126.0 MMtCO2e is from "field production," 17.1 MMtCO2e is from "processing," 43.8 MMtCO2e is from "transmission and storage," and 28.5 MMtCO2e is from "distribution." U.S. Environmental Protection Agency, *Inventory of U.S. Greenhouse Gas Emissions and Sinks: 1990-2010,* Washington, DC, EPA 430-R-12-001, April 15, 2012, http://www.epa.gov/climatechange/Downloads/ghgemissions/US-GHGInventory-2012-Main-Text.pdf. For a definition of carbon dioxide equivalent, see footnote 31.

[115] Global-warming potential (GWP) is a relative measure of how much heat a greenhouse gas traps in the atmosphere. It compares the amount of heat trapped by a certain mass of the gas in question to the amount of heat trapped by a similar mass of carbon dioxide. A GWP is calculated over a specific time interval, commonly 20, 100 or 500 years. See United Nations Framework Convention on Climate Change, "Global Warming Potentials" webpage, http://unfccc.int/ ghg_data/items/3825.php.

[116] As discussed in section "Measurement of Emissions."

[117] Robert W. Howarth, Renee Santoro, and Anthony Ingraffea, "Methane and the Greenhouse Gas Footprint of Natural Gas from Shale Formations," *Climatic Change*, vol. 106, no. 4 (June 2011), pp. 679-690, http://rd.springer.com/article/10.1007/s10584-011-0061-5.

[118] Lawrence M. Cathles III, Larry Brown, Milton Taam, and Andrew Hunter, "A Commentary on "Methane and the Greenhouse Gas Footprint of Natural Gas from Shale Formations" by R.W. Howarth, R. Santoro, and Anthony Ingraffea," *Climatic Change*, vol. 113, no. 2 (July 2012), pp. 525-535, http://rd.springer.com/article/10.1007/s10584- 011-0333-0.

[119] Nathan Hultman et al., "The Greenhouse Impact of Unconventional Gas for Electricity Generation," *Environmental Research Letters*, vol. 6, no. 4 (2011). http://iopscience.iop.org/1748-9326/6/4/044008.

[120] Mark Fulton, et al., "Comparing Life-Cycle Greenhouse Gas Emissions from Natural Gas and Coal," Deutsche Bank Group Climate Change Advisors and Worldwatch Institute, August 25, 2011, http://www.worldwatch.org/system/ files/pdf/Natural_Gas_LCA_Update_082511.pdf.

[121] Mohan Jiang, et al., "Life Cycle Greenhouse Gas Emissions of Marcellus Shale Gas," Environmental Research Letters, vol. 6 (July-September 2011), http://iopscience.iop.org/1748-9326/6/3/034014/fulltext/.

[122] Ramón A. Alvarez et al., "Greater Focus Needed on Methane Leakage from Natural Gas Infrastructure," *Proceedings of the National Academy of Sciences*, vol. 109, no. 17 (April 24, 2012), pp. 6435–6440, http://www.pnas.org/content/109/17/6435.

[123] Nathan Hultman et al., *op cit.*

[124] Gabrielle Pétron et al., "Hydrocarbon Emissions Characterization in the Colorado Front Range: A Pilot Study," *Journal of Geophysical Research*, vol. 117 (2012).

[125] Advanced Resources International Inc., *Estimate of Impacts of EPA Proposals to Reduce Air Emissions from Hydraulic Fracturing Operations,* prepared for the American Petroleum Institute, February 2012, http://www.api.org/~/ media/Files/Policy/Hydraulic_Fracturing/NSPS-OG-ARI-Impacts-of-EPA-Air-Rules-Final-Report.ashx.

[126] Thomas J. Pyle, "EPA's Flawed Rule Warrants Scrutiny," *National Journal,* April 23, 2012, http://energy.nationaljournal.com/2012/04/regulating-natural-gas-whats-t.php#2201556.

[127] U.S. Environmental Protection Agency, *Regulatory Impact Analysis: Final New Source Performance Standards and Amendments to the National Emissions Standards for*

Hazardous Air Pollutants for the Oil and Natural Gas Industry, April 2012, p. 3-6, http://www.epa.gov/ttnecas1/regdata/RIAs/oil_natural_gas_final_neshap_nsps_ria.pdf.

[128] Comments to "Oil and Natural Gas Sector: New Source Performance Standards and National Emission Standards for Hazardous Air Pollutants Reviews," submitted by Howard J. Feldman, Director, Regulatory and Scientific Affairs, American Petroleum Institute, November 30, 2011, accessed on http://www.regulations.gov; Jim Efstathiou Jr. and Mark Drajem," Gas Groups' Strategies Diverge in Opposing Obama Fracking Rules," *Bloomberg News*, March 26, 2012, http://www.bgov.com/news_item/tChIVzjpbV4O4xTaVVX-FA; Comments submitted by Devon Energy, Nov. 30, 2011, accessed on http://www.regulations.gov; David A. Simpson, P.E., *Review of the Regulatory Impact Analysis (RIA) for the Well Completions Portion of the Proposed New Source Performance Standards for the Oil and Natural Gas Industry,* Attachment G to comments by API, op cit.; Comments submitted by Amy Farrell, Vice President of Regulatory Affairs, America's Natural Gas Alliance (ANGA) and Bruce Thompson, President, American Exploration and Petroleum Council (AXPC), November 30, 2011, accessed on http://www.regulations.gov.

[129] Rich Heidorn Jr., *Fracking Emission Rules: EPA, Industry Miss Mark on Costs, Consequences,* Bloomberg Government, 2012, http://about.bgov.com/2012/07/19/fracking-emissions-rules-re-estimating-the-costs/. The report cites EIA figures of 24 trillion cubic feet of gas sold by U.S. natural gas producers in 2011, with a value of $72.5 billion at $3/Mcf.

[130] See comments submitted by Howard J. Feldman, API, *op cit.*

[131] See comments submitted by Sierra Club, *op cit.*

[132] 42 U.S.C. §7411 (a)(3).

[133] *Alabama Power Co. v. Costle,* 636 F.2d 323 (D.C. Cir. 1979).

[134] 45 *Federal Register* 52676, August 7, 1980.

[135] 45 *Federal Register* 52676, August 7, 1980.

[136] See "Withdrawal of Source Determinations for Oil and Gas Industries," memorandum from Gina McCarthy to Regional Administrators, September 22, 2009, withdrawing the 2007 EPA memo "Source Determinations for Oil and Gas Industries," memorandum from William L. Wehrum to Regional Administrators, January, 12, 2007.

[137] *Summit Petroleum Corp. v. EPA*, 6th Cir., Nos. 09-4348, 10-4572, 8/7/12.

[138] For example, the agency argues that the Summit ruling could actually lead to increased regulatory burdens under the NSR and PSD programs because permitting obligations are triggered by modifications at a source that result in net emission increases. Under the program, facilities can reduce emissions from another unit at that source to balance out emissions increases and avoid the trigger. EPA also says that relying on physical proximity alone will lead to "absurd results," saying that regulators as a result of the vacatur would have to consider a group of oil wells that are located near a series of gas wells and owned by the same operator to be one source although they emit different streams of pollutants.

[139] Representative Jack Fields (R-TX-08), Conference Report on S. 1630, Clean Air Act Amendments of 1990 (House of Representatives - October 26, 1990), *Congressional Record*, vol. 136, p. H12868.

[140] U.S. Environmental Protection Agency, "National Emission Standards for Hazardous Air Pollutants: Oil and Natural Gas Production and National Emission Standards for Hazardous Air Pollutants: Natural Gas Transmission and Storage," 64 *Federal Register* 32610, June 17, 1999.

In: Air Quality and the Natural Gas Industry ISBN: 978-1-62618-476-3
Editor: Jamie D. Zaleski

Chapter 2

CLEAN AIR ACT: A SUMMARY OF THE ACT AND ITS MAJOR REQUIREMENTS

James E. McCarthy, Claudia Copeland, Larry Parker and Linda-Jo Schierow

SUMMARY

This report summarizes the Clean Air Act and its major regulatory requirements.

The principal statute addressing air quality concerns, the Clean Air Act was first enacted in 1955, with major revisions in 1970, 1977, and 1990. The Act requires EPA to set health-based standards for ambient air quality, sets deadlines for the achievement of those standards by state and local governments, and requires EPA to set national emission standards for large or ubiquitous sources of air pollution, including motor vehicles, power plants, and other industrial sources. In addition, the Act mandates emission controls for sources of 187 hazardous air pollutants, establishes a cap-and-trade program to limit acid rain, requires the prevention of significant deterioration of air quality in areas with clean air, requires a program to restore visibility impaired by regional haze in national parks and wilderness areas, and implements the Montreal Protocol to phase out most ozone-depleting chemicals.

This report describes the Act's major provisions and provides tables listing all major amendments, with the year of enactment and Public Law number, and cross-referencing sections of the Act with the major U.S. Code sections of the codified statute.

INTRODUCTION

The authorities and responsibilities of the Environmental Protection Agency (EPA) derive primarily from a dozen major environmental statutes. This report provides a concise summary of one of those statutes, the Clean Air Act. It provides a very brief history of federal involvement in air quality regulation and of the provisions added by legislation in 1970, 1977, and 1990; it explains major authorities contained in the Act; it defines key terms; and it lists references for more detailed information on the Act and its implementation.

While this report attempts to present the essence of the Act, it is necessarily incomplete. Many details and secondary provisions are omitted. In addition, the report describes the statute largely without discussing its implementation. Statutory deadlines to control emissions and achieve particular mandates have often been missed as a result of delayed standard-setting by EPA, delayed action on implementation by states and local governments, or law suits brought by interested parties.

OVERVIEW

The Clean Air Act, codified as 42 U.S.C. 7401 *et seq.*, seeks to protect human health and the environment from emissions that pollute ambient, or outdoor, air. It requires the Environmental Protection Agency to establish minimum national standards for air quality, and assigns primary responsibility to the states to assure compliance with the standards. Areas not meeting the standards, referred to as "nonattainment areas," are required to implement specified air pollution control measures. The Act establishes federal standards for mobile sources of air pollution and their fuels and for sources of 187 hazardous air pollutants, and it establishes a cap-and-trade program for the emissions that cause acid rain. It establishes a comprehensive permit system for all major sources of air pollution. It also addresses the prevention of pollution in areas with clean air and protection of the stratospheric ozone layer.

Like many other programs administered by the Environmental Protection Agency, federal efforts to control air pollution have gone through several phases, beginning with information collection, research, and technical assistance, before being strengthened to establish federal standards and enforcement. Federal legislation addressing air pollution was first passed in 1955, prior to which air pollution was the exclusive responsibility of state and local levels of government.

The federal role was strengthened in subsequent amendments, notably the Clean Air Act Amendments of 1970, 1977, and 1990. The 1970 amendments established the procedures under which EPA sets national standards for air quality, required a 90% reduction in emissions from new automobiles by 1975, established a program to require the best available control technology at major new sources of air pollution, established a program to regulate air toxics, and greatly strengthened federal enforcement authority. The 1977 amendments adjusted the auto emission standards, extended deadlines for the attainment of air quality standards, and added the Prevention of Significant Deterioration program to protect air cleaner than national standards.

Changes to the Act in 1990 included provisions to (1) classify most nonattainment areas according to the extent to which they exceed the standard, tailoring deadlines, planning, and controls to each area's status; (2) tighten auto and other mobile source emission standards; (3) require reformulated and alternative fuels in the most polluted areas; (4) revise the air toxics section, establishing a new program of technology-based standards and addressing the problem of sudden, catastrophic releases of air toxics; (5) establish an acid rain control program, with a marketable allowance scheme to provide flexibility in implementation; (6) require a state-run permit program for the operation of major sources of air pollutants; (7) implement the Montreal Protocol to phase out most ozone-depleting chemicals; and (8) update the enforcement provisions so that they parallel those in other pollution control acts, including authority for EPA to assess administrative penalties.

The 1990 amendments also authorized appropriations for clean air programs through FY1998. The Act has not been reauthorized since then. House rules require enactment of an authorization before an appropriation bill can be considered; but this requirement can be waived and frequently has been. Thus, while authorization of appropriations in the Clean Air Act (and most other environmental statutes) has expired, programs have continued and have been funded. The Act's other legal authorities, to issue and enforce regulations, are, for the most part, permanent and are not affected by the lack of authorization.

The remainder of this report describes major programs required by the Act, with an emphasis on the changes established by the 1990 amendments.

Table 1. Clean Air Act and Amendments (codified generally as 42 U.S.C. 7401-7671)

Year	Act	Public Law Number
1955	Air Pollution Control Act	P.L. 84-159
1959	Reauthorization	P.L. 86-35 3
1960	Motor vehicle exhaust study	P.L. 86-493
1963	Clean Air Act Amendments	P.L. 88-206
1965	Motor Vehicle Air Pollution Control Act	P.L. 89-272, Title I
1966	Clean Air Act Amendments of 1966	P.L. 89-675
1967	Air Quality Act of 1967	P.L. 90-148
	National Air Emission Standards Act	
1970	Clean Air Act Amendments of 1970	P.L. 91-604
1973	Reauthorization	P.L. 93-13
1974	Energy Supply and Environmental Coordination Act of 1974	P.L. 93-319
1977	Clean Air Act Amendments of 1977	P.L. 95-95
1980	Acid Precipitation Act of 1980	P.L. 96-294, Title VII
1981	Steel Industry Compliance Extension Act of 1981	P.L. 97-23
1987	Clean Air Act 8-month Extension	P.L. 100-202
1990	Clean Air Act Amendments of 1990	P.L. 101-549
1991	Technical correction to list of hazardous air pollutants	P.L. 102-187
1995-96	Relatively minor laws amending the Act	P.L. 104-6, 59, 70, 260
1998	Amended § 604 re methyl bromide	P.L. 105-277, Section 764
1998	Border Smog Reduction Act of 1998	P.L. 105-286
1999	Chemical Safety Information, Site Security and Fuels Regulatory Relief Act	P.L. 106-40
2004	Amendments to §209 re small engines	P.L. 108-199, Division G, Title IV, Section 428
2005	Energy Policy Act of 2005 (amended §211 re fuels)	P.L. 109-58
2007	Energy Independence and Security Act of 2007 (amended §211 re fuels)	P.L. 110-140

NATIONAL AMBIENT AIR QUALITY STANDARDS

In section 109, the Act requires EPA to establish National Ambient Air Quality Standards (NAAQS) for air pollutants that endanger public health or welfare, in the Administrator's judgment, and whose presence in ambient air results from numerous or diverse sources. The NAAQS must be designed to protect public health and welfare with an adequate margin of safety. Using this authority, EPA has promulgated NAAQS for six air pollutants: sulfur dioxide (SO_2), particulate matter (PM2.5 and PM10), nitrogen dioxide (NO_2), carbon monoxide (CO), ozone,[1] and lead. The Act requires EPA to review the scientific data upon which the standards are based, and revise the standards, if necessary, every five years. More often than not, however, EPA has taken more than five years in reviewing and revising the standards.

Originally, the Act required that the NAAQS be attained by 1977 at the latest, but the states experienced widespread difficulty in complying with this deadline. As a result, the deadlines for achieving NAAQS have been extended several times. Under the 1990 amendments, most areas not in attainment with NAAQS must meet special compliance schedules, staggered according to the severity of an area's air pollution problem. The amendments also established specific requirements for each nonattainment category, as described below.

STATE IMPLEMENTATION PLANS

While the Act authorizes the EPA to set NAAQS, the states are responsible for establishing procedures to attain and maintain the standards. Under Section 110 of the Act, the states adopt plans, known as State Implementation Plans (SIPs), and submit them to EPA to ensure that they are adequate to meet statutory requirements.

SIPs are based on emission inventories and computer models to determine whether air quality violations will occur. If these data show that standards would be exceeded, the state must impose additional controls on existing sources to ensure that emissions do not cause "exceedances" of the standards. Proposed new and modified sources must obtain state construction permits in which the applicant shows how the anticipated emissions will not exceed allowable limits. In nonattainment areas, emissions from new or modified sources must also be offset by reductions in emissions from existing sources.

The 1990 amendments require EPA to impose sanctions in areas which fail to submit a SIP, fail to submit an adequate SIP, or fail to implement a SIP: unless the state corrects such failures, a 2-to-1 emissions offset for the construction of new polluting sources is imposed 18 months after notification to the state, and a ban on most new federal highway grants is imposed six months later. An additional ban on air quality grants is discretionary.

Ultimately, a Federal Implementation Plan may be imposed if the state fails to submit or implement an adequate SIP.

The amendments also require that, in nonattainment areas, no federal permits or financial assistance may be granted for activities that do not "conform" to a State Implementation Plan. This requirement can cause a temporary suspension in funding for most new highway and transit projects if an area fails to demonstrate that the emissions caused by such projects are consistent with attainment and maintenance of ambient air quality standards. Demonstrating conformity of transportation plans and SIPs is required in nonattainment areas whenever new plans are submitted.

Nonattainment Requirements

In a major departure from the prior law, the 1990 Clean Air Act Amendments grouped most nonattainment areas into classifications based on the extent to which the NAAQS was exceeded, and established specific pollution controls and attainment dates for each classification. These requirements are described here as spelled out in Sections 181-193 of the Act.[2]

Nonattainment areas are classified on the basis of a "design value," which is derived from the pollutant concentration (in parts per million or micrograms per cubic meter) recorded by air quality monitoring devices. The design value for the 1-hour ozone standard was the fourth highest hourly reading measured during the most recent three-year period. Using these design values, the Act created five classes of ozone nonattainment, as shown in **Table 2**. Only Los Angeles fell into the "extreme" class, but 97 other areas were classified in one of the other four ozone categories. A simpler classification system established moderate and serious nonattainment areas for carbon monoxide and particulate matter with correspondingly more stringent control requirements for the more polluted class.

As shown in the table, the deadlines for attainment for ozone nonattainment areas stretched from 1993 to 2010, depending on the severity of the problem. (Under the 8-hour ozone standard, which replaced the 1-hour standard in 2004, these deadlines are changed to 2007 to 2021.) For carbon monoxide, the attainment date for moderate areas was December 31, 1995, and for serious areas, December 31, 2000. For particulate matter, the deadline for areas designated moderate nonattainment as of 1990 was December 31, 1994; for those areas subsequently designated as moderate, the deadline is six years after designation. For serious areas, the respective deadlines are December 31, 2001 or 10 years after designation.

Table 2. Ozone Nonattainment Classifications

Class	Marginal	Moderate	Serious	Severe	Extreme
Deadline	1993	1996	1999	2005-2007[a]	2010
Areas[b]	42 areas	32 areas	14 areas	9 areas	1 area
Design Value	0.121 ppm-0.138 ppm	0.138 ppm-0.160 ppm	0.160 ppm-0.180 ppm	0.180 ppm-0.280 ppm	>0.280 ppm

[a] Areas with a 1988 design value between 0.190 and 0.280 ppm have 17 years to attain; others have 15 years.

[b] Number of areas in each category as of the date of enactment.

Requirements for Ozone Nonattainment Areas

Although areas with more severe air pollution problems have a longer time to meet the standards, more stringent control requirements are imposed in areas with worse pollution. A summary of the primary ozone control requirements for each nonattainment category follows.

Marginal Areas

- Inventory emissions sources (to be updated every three years).
- Require 1.1 to 1 offsets (i.e., new major emission sources of volatile organic compounds [VOCs] must reduce VOC emissions from existing facilities in the area by 10% more than the emissions of the new facility).
- Impose reasonably available control technology (RACT) on all major sources emitting more than 100 tons per year for the nine industrial categories where EPA had already issued control technique guidelines describing RACT prior to 1990.

Moderate Areas

- Meet all requirements for marginal areas.
- Impose a 15% reduction in VOC emissions in six years.
- Adopt a basic vehicle inspection and maintenance program.
- Impose RACT on all major sources emitting more than 100 tons per year for all additional industrial categories where EPA will issue control technique guidelines describing RACT.

- Require vapor recovery at gas stations selling more than 10,000 gallons per month.
- Require 1.15 to 1 offsets.

Serious Areas

- Meet all requirements for moderate areas.
- Reduce definition of a major source of VOCs from emissions of 100 tons per year to 50 tons per year for the purpose of imposing RACT.
- Reduce VOCs 3% annually for years 7 to 9 after the 15% reduction already required by year 6.
- Improve monitoring.
- Adopt an enhanced vehicle inspection and maintenance program.
- Require fleet vehicles to use clean alternative fuels.
- Adopt transportation control measures if the number of vehicle miles traveled in the area is greater than expected.
- Require 1.2 to 1 offsets.
- Adopt contingency measures if the area does not meet required VOC reductions.

Severe Areas

- Meet all requirements for serious areas.
- Reduce definition of a major source of VOCs from emissions of 50 tons per year to 25 tons per year for the purpose of imposing RACT.
- Adopt specified transportation control measures.
- Implement a reformulated gasoline program.
- Require 1.3 to 1 offsets.
- Impose $5,000 per ton penalties on major sources if the area does not meet required reductions.

Extreme Areas

- Meet all requirements for severe areas.
- Reduce definition of a major source of VOCs from emissions of 25 tons per year to 10 tons per year for the purpose of imposing RACT.
- Require clean fuels or advanced control technology for boilers emitting more than 25 tons per year of NO_x.
- Require 1.5 to 1 offsets.

As noted, EPA promulgated a new, 8-hour ozone standard in July 1997. Following extensive court challenges, the agency designated nonattainment areas for the new standard on April 30, 2004. State Implementation Plans were required to be submitted in 2007.

Requirements for Carbon Monoxide Nonattainment Areas

As with ozone nonattainment areas, carbon monoxide (CO) nonattainment areas are subjected to specified control requirements, with more stringent requirements in Serious nonattainment areas. A summary of the primary CO control requirements for each nonattainment category follows.

Moderate Areas

- Conduct an inventory of emissions sources.
- Forecast total vehicle miles traveled in the area.
- Adopt an enhanced vehicle inspection and maintenance program.
- Demonstrate annual improvements sufficient to attain the standard.

Serious Areas

- Adopt specified transportation control measures.
- Implement an oxygenated fuels program for all vehicles in the area.
- Reduce definition of a major source of CO from emissions of 100 tons per year to 50 tons per year if stationary sources contribute significantly to the CO problem.

Serious areas failing to attain the standard by the deadline have to revise their SIP and demonstrate reductions of 5% per year until the standard is attained.

Requirements for Particulate Nonattainment Areas

Particulate (PM10) nonattainment areas are also subject to specified control requirements. These are:

Moderate Areas

- Require permits for new and modified major stationary sources of PM10.
- Impose reasonably available control measures (RACM).

Serious Areas

- Impose best available control measures (BACM).
- Reduce definition of a major source of PM_{10} from 100 tons per year to 70 tons per year.

In July 1997, EPA promulgated new standards for fine particulates $(PM_{2.5})$. The $PM_{2.5}$ standards were also subject to court challenges. The absence of a monitoring network capable of measuring the pollutant delayed implementation as well. Nonattainment areas for $PM_{2.5}$ were designated on April 14, 2005. States had three years subsequent to designation to submit State Implementation Plans. Revisions to the NAAQS promulgated in October 2006 strengthened the $PM_{2.5}$ standard.

TRANSPORTED AIR POLLUTION

Meeting the nation's clean air standards can be complicated as air pollution is no respecter of political boundaries or subdivisions. This problem of transported air pollutants has come into particular focus as states and EPA attempt to develop effective compliance strategies to achieve both the ozone and the PM2.5 NAAQS. Under Section 110(a)(2)(D), SIPs must include adequate provisions to prevent sources within that state from contributing significantly to nonattainment in one or more downwind states.

If EPA finds a SIP inadequate to achieve a NAAQS, it must require the affected state to submit a revised SIP that includes sufficient measures to bring that state into compliance. This is known as a "SIP Call." The 1990 Clean Air Act amendments provided EPA and the states with new tools to address the transport problem through this provision. One of those tools is Section 176A, a provision that permits the EPA, either on its own or by petition from any state, to establish a transport region to address regional pollution problems contributing to violations of a primary NAAQS. A commission of EPA and state officials is constituted to make recommendations to EPA on appropriate mitigation strategies. Based on the commission's findings and recommendations, EPA is then required under section 1 10(k)(5) to notify affected states of inadequacies in their current state implementation plans and to establish deadlines (not to exceed 18 months) for submitting necessary revisions (i.e., a SIP call). Besides authorizing administratively-created transport regions, the 1990 amendments statutorily created an Ozone Transport Region (OTR) in the Northeast. This provision (Section 184 of the act) required specific additional controls for all areas (not only nonattainment areas) in that region, and established the Ozone Transport Commission for the purpose of recommending to EPA regionwide controls affecting all areas in the region.

The transport issue may also be addressed by affected downwind states through a Section 126 petition. As amended by the 1990 Clean Air Act amendments, under Section 126(b) any state or political subdivision may petition EPA for a finding that a major source or group of stationary sources located in another state is emitting pollutants that "significantly contribute" to the nonattainment of a NAAQS by their state. EPA is to respond to the petition within 60 days. If the petition is granted, the offending sources must cease operations within three months unless the sources comply with emission controls and the compliance schedules as determined by EPA to bring them into compliance with the section. Section 126 has rarely been used, although it has proven useful to EPA in some cases as backup authority where there might be challenges to a SIP call.

Emission Standards for Mobile Sources

Title II of the Clean Air Act has required emission standards for automobiles since 1968. The 1990 amendments significantly tightened these standards: for cars, the hydrocarbon standard was reduced by 40% and the nitrogen oxides (NO_x) standard by 50%. These standards—referred to as "Tier 1" standards—were phased in over the 1994-1996 model years.

The amendments envisioned a further set of reductions ("Tier 2" standards), but not before model year 2004. For Tier 2 standards to be promulgated, the agency was first required to report to Congress concerning the need for further emission reductions, the availability of technology to achieve such reductions, and the cost-effectiveness of such controls compared to other means of attaining air quality standards. EPA submitted this report to Congress in August 1998, concluding that further emission reductions were needed and that technology to achieve such reductions was available and cost-effective. Tier 2 standards, requiring emission reductions of 77% to 95% from cars and light trucks were promulgated in February 2000, and were phased in over the 2004-2009 model years. To facilitate the use of more effective emission controls, the standards also require a more than 90% reduction in the sulfur content of gasoline, beginning in 2004.

The 1990 amendments also required that oxygenated gasoline, designed to reduce emissions of carbon monoxide, be sold in the worst CO nonattainment areas and that "reformulated" gasoline (RFG), designed to reduce emissions of volatile organic compounds and toxic air pollutants, be sold in the nine worst ozone nonattainment areas (Los Angeles, San Diego, Houston, Baltimore, Philadelphia, New York, Hartford, Chicago, and Milwaukee); metropolitan Washington, DC; and four areas in California were added to the mandatory list later. Other ozone nonattainment areas can opt in to the RFG program; as of 2006, additional areas in 11 states had done so.

The fuels provisions were modified by the Energy Policy Act of 2005 (EPACT), removing the requirement that RFG contain oxygenates. Instead, EPACT required the use of increasing amounts of renewable fuel, most likely to be ethanol, in motor fuels, beginning in 2006. The Energy Independence and Security Act of 2007 further strengthened the renewable fuel requirements.

Use of alternative fuels and development of cleaner engines was also to be stimulated by the Clean-Fuel Fleet Program. In all of the most seriously polluted ozone and CO nonattainment areas, centrally fueled fleets of 10 or more passenger cars and light-duty trucks must purchase at least 30% clean-fuel vehicles when they add new vehicles to existing fleets, starting in 1999. (The Act originally required the program to begin in 1998, but the start was delayed by a year.) The percentage rose to 50% in 2000 and 70% in 2001. Heavy-duty fleets are required to purchase at least 50% clean-fuel vehicles annually. A clean fuel vehicle is one which meets Low Emission Vehicle (LEV) standards and operates on reformulated gasoline, reformulated diesel, methanol, ethanol, natural gas, liquefied petroleum gas, hydrogen, or electricity.

In addition to the above program, California's Zero Emission Vehicle (ZEV) program also is intended to promote the development of alternative fuels and vehicles. Section 209(b) of the Clean Air Act allows the EPA Administrator to grant California the authority to develop its own vehicle emissions standards if those standards are at least as stringent as the federal standards and if the state demonstrates that it needs the standards to meet compelling and extraordinary conditions. In addition to setting more stringent standards for all vehicles, California used this authority to establish a program requiring auto manufacturers to sell ZEVs (electric or hydrogen fuel cell vehicles) in the state beginning in 2003. This program has been substantially modified since it was enacted, and now allows credit for hybrid and partial ZEV vehicles in addition to true ZEVs, but it has served as an incubator for lower emission technologies since its adoption. Section 177 of the Act allows other states to adopt California's stricter standards: at least ten states (Connecticut, Maine, Massachusetts, New Jersey, New York, Oregon, Pennsylvania, Rhode Island, Vermont, and Washington) have already adopted them or are in the process of doing so.

The 1990 amendments also imposed tighter requirements on certification (an auto's useful life is defined as 100,000 miles instead of the earlier 50,000 miles), on emissions allowed during refueling, on low temperature CO emissions, on in-use performance over time, and on warranties for the most expensive emission control components (8 years/80,000 miles for the catalytic converter, electronic emissions control unit, and onboard emissions diagnostic unit). Regulations were also extended to include nonroad fuels and engines.

Standards for trucks and buses using diesel engines were also strengthened. The 1990 amendments required new urban buses to reduce emissions of diesel particulates 92% by 1996, and all other heavy-duty diesel engines to achieve an 83% reduction by the same year. NO_x emissions must also be reduced, 33% by 1998. Authority to further strengthen these standards led to promulgation in January 2001 of new emission standards requiring a further 90%-95% reduction in emissions phased in over the 2007-2010 model years, and a reduction of 97% in the allowable amount of sulfur in highway diesel fuel. These regulations were followed in May 2004 by similar requirements for nonroad diesel equipment, which will be phased in between 2007 and 2015.

Hazardous Air Pollutants[3]

Completely rewritten by the Clean Air Act Amendments of 1990, Section 112 of the Act establishes programs for protecting public health and the environment from exposure to toxic air pollutants. As revised by the 1990 amendments, the section contains four major provisions: Maximum Achievable Control Technology (MACT) requirements; health-based standards; standards for stationary "area sources" (small, but numerous sources, such as gas stations or dry cleaners, that collectively emit significant quantities of hazardous pollutants); and requirements for the prevention of catastrophic releases.

First, EPA is to establish technology-based emission standards, called MACT standards, for sources of 187 pollutants listed in the legislation, and to specify categories of sources subject to the emission standards.[4] EPA is to revise the standards periodically (at least every eight years). EPA can, on its own initiative or in response to a petition, add or delete substances or source categories from the lists.

Section 112 establishes a presumption in favor of regulation for the designated chemicals; it requires regulation of a designated pollutant unless EPA or a petitioner is able to show "that there is adequate data on the health and environmental effects of the substance to determine that emissions, ambient concentrations, bioaccumulation or deposition of the substance may not reasonably be anticipated to cause any adverse effects to human health or adverse environmental effects."

EPA is required to set standards for sources of the listed pollutants that achieve "the maximum degree of reduction in emissions" taking into account cost and other non-air-quality factors. These MACT standards for new sources "shall not be less stringent than the most stringent emissions level that is achieved in practice by the best controlled similar source." The standards for existing sources may be less stringent than those for new sources, but must be no less stringent than the emission limitations achieved by either the best performing 12% of existing sources (if there are more than 30 such sources in the category or subcategory) or the best performing 5 similar sources (if there are fewer than 30). Existing sources are given three years following promulgation of standards to achieve compliance, with a possible one-year extension; additional extensions may be available for special circumstances or for certain categories of sources. Existing sources that achieve voluntary early emissions reductions will receive a six-year extension for compliance with MACT.

The second major provision of Section 112 directs EPA to set health-based standards to address situations in which a significant residual risk of adverse health effects or a threat of adverse environmental effects remains after installation of MACT. This provision requires that EPA, after consultation with the Surgeon General of the United States, submit a report to Congress on the public health significance of residual risks, and recommend legislation regarding such risks. If Congress does not legislate in response to EPA's recommendations, then EPA is required to issue standards for categories of sources of hazardous air pollutants as necessary to protect the public health with an ample margin of safety or to prevent an adverse environmental effect. A residual risk standard is required for any source emitting a cancer-causing pollutant that poses an added risk to the most exposed person of more than one-in-a-million. Residual risk standards are due eight years after promulgation of MACT for the affected source category.

Existing sources have 90 days to comply with a residual risk standard, with a possible two-year extension. In general, residual risk standards do not apply to area sources.

The law directed EPA to contract with the National Academy of Sciences (NAS) for a study of risk assessment methodology, and created a Risk Assessment and Management Commission to investigate and report on policy implications and appropriate uses of risk assessment and risk management. In 1994 NAS published its report, *Science and Judgment in Risk Assessment*. The Commission study, *Framework for Environmental Health Risk Management*, was released in 1997.

Third, in addition to the technology-based and health-based programs for major sources of hazardous air pollution, EPA is to establish standards for stationary "area sources" determined to present a threat of adverse effects to human health or the environment. The provision requires EPA to regulate the stationary area sources responsible for 90% of the emissions of the 30 hazardous air pollutants that present the greatest risk to public health in the largest number of urban areas. In setting the standard, EPA can impose less stringent "generally available" control technologies, rather than MACT.

Finally, Section 112 addresses prevention of sudden, catastrophic releases of air toxics by establishing an independent Chemical Safety and Hazard Investigation Board. The Board is responsible for investigating accidents involving releases of hazardous substances, conducting studies, and preparing reports on the handling of toxic materials and measures to reduce the risk of accidents.

EPA is also directed to issue prevention, detection, and correction requirements for catastrophic releases of air toxics by major sources. Section 112(r) requires owners and operators to prepare risk management plans including hazard assessments, measures to prevent releases, and a response program.

New Source Performance Standards[5]

Section 111 of the Act requires EPA to establish nationally uniform, technology-based standards (called New Source Performance Standards, or NSPS) for categories of new industrial facilities. These standards accomplish two goals: first, they establish a consistent baseline for pollution control that competing firms must meet, and thereby remove any incentive for states or communities to weaken air pollution standards in order to attract polluting industry; and second, they preserve clean air to accommodate future growth, as well as for its own benefits.

NSPS establish maximum emission levels for new major stationary sources—powerplants, steel mills, and smelters, for example—with the emission levels determined by the best "adequately demonstrated" continuous control technology available, taking costs into account. EPA must regularly revise and update NSPS applicable to designated sources as new technology becomes available, since the goal is to prevent new pollution problems from developing and to force the installation of new control technology.

The standards also apply to modifications of existing facilities, through a process called New Source Review (NSR). The law's ambiguity regarding what constitutes a modification (subject to NSR) as opposed to routine maintenance of a facility has led to litigation, with EPA proposing in recent years to modify its interpretation of the requirements of this section.

SOLID WASTE INCINERATORS

Prior to 1990, solid waste incinerators, which emit a wide range of pollutants, were subject to varying degrees of state and federal regulation depending on their size, age, and the type of waste burned. In a new Section 129, the 1990 amendments established more consistent federal requirements specifying that emissions of 10 categories of pollutants be regulated at new and existing incinerators burning municipal solid waste, medical waste, and commercial and industrial waste. The amendments also established emissions monitoring and operator training requirements.

PREVENTION OF SIGNIFICANT DETERIORATION / REGIONAL HAZE[6]

Sections 160-169 of the act establish requirements for the prevention of significant deterioration of air quality (PSD). The PSD program reflects the principle that areas where air quality is better than that required by NAAQS should be protected from significant new air pollution even if NAAQS would not be violated.

The Act divides clean air areas into three classes, and specifies the increments of SO_2 and particulate pollution allowed in each. Class I areas include international and national parks, wilderness and other pristine areas; allowable increments of new pollution are very small. Class II areas include all attainment and not classifiable areas, not designated as Class I; allowable increments of new pollution are modest. Class III represents selected areas that states may designate for development; allowable increments of new pollution are large (but not exceeding NAAQS). Through an elaborate hearing and review process, a state can have regions redesignated from Class II to Class III (although none have yet been so redesignated).

While the 1977 amendments only stipulated PSD standards for two pollutants, SO_2 and particulates, EPA is supposed to establish standards for other criteria pollutants. Thus far, only one of the other four (NO_2) has been addressed: the agency promulgated standards for NO_2 in 1988.

Newly constructed polluting sources in PSD areas must install best available control technology (BACT) that may be more strict than that required by NSPS. The justifications of the policy are that it protects air quality, provides an added margin of health protection, preserves clean air for future development, and prevents firms from gaining a competitive edge by "shopping" for clean air to pollute.

In Sections 1 69A and B, the Act also sets a national goal of preventing and remedying impairment of visibility in national parks and wilderness areas, and requires EPA to promulgate regulations to assure reasonable progress toward that goal. In the 1990 Amendments, Congress strengthened these provisions, which had not been implemented.

The amendments required EPA to establish a Grand Canyon Visibility Transport Commission, composed of Governors from each state in the affected region, an EPA designee, and a representative of each of the national parks or wilderness areas in the region. Other visibility transport commissions can be established upon EPA's discretion or upon petition from at least two states. Within 18 months of receiving a report from one of these commissions, EPA is required to promulgate regulations to assure reasonable progress toward the visibility goal, including requirements that states update their State Implementation Plans to contain emission limits, schedules of compliance, and other measures necessary to make reasonable progress. Specifically mentioned is a requirement that states impose Best Available Retrofit Technology on existing sources of emissions impairing visibility.

The Grand Canyon Commission delivered a set of recommendations to EPA in June 1996, and the agency subsequently promulgated a "regional haze" program applicable to all 50 states under this authority.

ACID DEPOSITION CONTROL[7]

The Clean Air Act Amendments of 1990 added an acid deposition control program (Title IV) to the Act. It set goals for the year 2000 of reducing annual SO_2 emissions by 10 million tons from 1980 levels and reducing annual NO_x emissions by 2 million tons, also from 1980 levels.

The SO_2 reductions were imposed in two steps. Under Phase 1, owners/operators of 111 electric generating facilities listed in the law that are larger than 100 megawatts had to meet tonnage emission limitations by January 1, 1995. This would reduce SO_2 emission by about 3.5 million tons. Phase 2 included facilities larger than 75 megawatts, with a deadline of January 1, 2000. Compliance has been 100%.

To introduce some flexibility in the distribution and timing of reductions, the Act creates a comprehensive permit and emissions allowance system. An allowance is a limited authorization to emit a ton of SO_2. Issued by EPA, the allowances would be allocated to Phase 1 and Phase 2 units in accordance with baseline emissions estimates. Powerplants which commence operation after November 15, 1990 would not receive any allowances. These new units would have to obtain allowances (offsets) from holders of existing allowances. Allowances may be traded nationally during either phase. The law also permitted industrial sources and powerplants to sell allowances to utility systems under regulations developed by EPA. Allowances may be banked by a utility for future use or sale.

The Act provided for two types of sales to improve the liquidity of the allowance system and to ensure the availability of allowances for utilities and independent power producers who need them. First, a special reserve fund consisting of 2.8% of Phase 1 and Phase 2 allowance allocations has been set aside for sale. Allowances from this fund (25,000 annually from 1993-1999 and 50,000 thereafter) are sold at a fixed price of $1,500 an allowance. Independent power producers have guaranteed rights to these allowances under certain conditions. Second, an annual, open auction sold allowances (150,000 from 1993-1995, and 250,000 from 1996-1999) with no minimum price. Utilities with excess allowances may have them auctioned off at this auction, and any person may buy allowances.

The Act essentially caps SO_2 emissions at individual existing sources through a tonnage limitation, and at future plants through the allowance system. First, emissions from most existing sources are capped at a specified emission rate times an historic baseline level. Second, for plants commencing operation after November 15, 1990, emissions must be completely offset with additional reductions at existing facilities beginning after Phase 2 compliance. However, as noted above, the law provides some allowances to future powerplants which meet certain criteria. The utility SO_2 emission cap was set at 8.9 million tons, with some exceptions.

The Act provides that if an affected unit does not have sufficient allowances to cover its emissions, it is subject to an excess emission penalty of \$2,000 per ton of SO_2 and required to reduce an additional ton of SO_2 the next year for each ton of excess pollutant emitted.

The Act also requires EPA to inventory industrial emissions of SO_2 and to report every five years, beginning in 1995. If the inventory shows that industrial emissions may reach levels above 5.60 million tons per year, then EPA is to take action under the Act to ensure that the 5.60 million ton cap is not exceeded.

The Act requires EPA to set specific NO_x emission rate limitations—0.45 lb. per million Btu for tangentially-fired boilers and 0.50 lb. per million Btu for wall-fired boilers—unless those rates can not be achieved by low-NO_x burner technology. Tangentially and wall-fired boilers affected by Phase 1 SO_2 controls must also meet NO_x requirements. EPA was to set emission limitations for other types of boilers by 1997 based on low-NO_x burner costs, which EPA did. In addition, EPA was to propose and promulgate a revised new source performance standard for NO_x from fossil fuel steam generating units, which EPA also did, in 1998.

PERMITS[8]

The Clean Air Act Amendments of 1990 added a Title V to the Act which requires states to administer a comprehensive permit program for the operation of sources emitting air pollutants. These requirements are modeled after similar provisions in the Clean Water Act. Previously, the Clean Air Act contained limited provision for permits, requiring only new or modified major stationary sources to obtain construction permits (under Section 165 of the Act).

Sources subject to the permit requirements generally include major sources that emit or have the potential to emit 100 tons per year of any regulated pollutant, plus stationary and area sources that emit or have potential to emit lesser specified amounts of hazardous air pollutants. However, in nonattainment areas, the permit requirements also include sources which emit as little as 50, 25, or 10 tons per year of VOCs, depending on the severity of the region's nonattainment status (serious, severe, or extreme).

States were required to develop permit programs and to submit those programs for EPA approval by November 15, 1993. EPA had one year to approve or disapprove a state's submission in whole or in part. After the effective date of a state plan, sources had 12 months to submit an actual permit application.

States are to collect annual fees from sources sufficient to cover the "reasonable costs" of administering the permit program, with revenues to be used to support the agency's air pollution control program. The fee must be at least $25 per ton of regulated pollutants (excluding carbon monoxide). Permitting authorities have discretion not to collect fees on emissions in excess of 4,000 tons per year and may collect other fee amounts, if appropriate.

The permit states how much of which air pollutants a source is allowed to emit. As a part of the permit process, a source must prepare a compliance plan and certify compliance.

The term of permits is limited to no more than five years; sources are required to renew permits at that time. State permit authorities must notify contiguous states of permit applications that may affect them; the application and any comments of contiguous states must be forwarded to EPA for review. EPA can veto a permit; however, this authority is essentially limited to major permit changes. EPA review need not include permits which simply codify elements of a state's overall clean air plan, and EPA has discretion to not review permits for small sources.

Holding a permit to some extent shields a source from enforcement actions: the Act provides that a source cannot be held in violation if it is complying with explicit requirements addressed in a permit, or if the state finds that certain provisions do not apply to that source.

ENFORCEMENT

Section 113 of the Act, which was also strengthened by the 1990 amendments, covers enforcement. The section establishes federal authority to issue agency and court orders requiring compliance and to impose penalties for violations of Act requirements.

Section 114 authorizes EPA to require sources to submit reports, monitor emissions, and certify compliance with the Act's requirements, and authorizes EPA personnel to conduct inspections.

Like most federal environmental statutes, the Clean Air Act is enforced primarily by states or local governments; they issue most permits, monitor compliance, and conduct the majority of inspections.

The federal government functions as a backstop, with authority to review state actions. The agency may act independently or may file its own enforcement action in cases where it concludes that a state's response was inadequate.

The Act also provides for citizen suits both against persons (including corporations or government agencies) alleged to have violated emissions standards or permit requirements, and against EPA in cases where the Administrator has failed to perform an action that is not discretionary under the Act.

Citizen groups have often used the latter provision to compel the Administrator to promulgate regulations required by the statute.

The 1990 Amendments elevated penalties for some knowing violations from misdemeanors to felonies; removed the ability of a source to avoid an enforcement order or civil penalty by ceasing a violation within 60 days of notice; gave authority to EPA to assess administrative penalties; and authorized $10,000 awards to persons supplying information leading to convictions under the Act.

Stratospheric Ozone Protection[9]

Title VI of the 1990 Clean Air Act Amendments represents the United States' primary response on the domestic front to the ozone depletion issue. It also implements the U.S. international responsibilities under the Montreal Protocol on Substances that Deplete the Ozone Layer (and its amendments). Indeed, Section 606(a)(3) provides that the Environmental Protection Agency shall adjust phase-out schedules for ozone-depleting substances in accordance with any future changes in Montreal Protocol schedules.

As a result, the phase-out schedules contained in Title VI for various ozone-depleting compounds have now been superseded by subsequent amendments to the Montreal Protocol.

Since passage of Title VI, depleting substances such as CFCs, methyl chloroform, carbon tetrachloride, and halons (referred to as Class 1 substances) have been phased out by industrial countries, including the United States. New uses of hydrochlorofluorocarbons (HCFCs) (called Class 2 substances under Title VI) are banned beginning January 1, 2015, unless the HCFCs are recycled, used as a feedstock, or used as a refrigerant for appliances manufactured prior to January 1, 2020.

Production of HCFCs is to be frozen January 1, 2015 and phased out by January 1, 2030. Exemptions consistent with the Montreal Protocol are allowed.

The EPA is required to add any substance with an ozone depletion potential (ODP) of 0.2 or greater to the list of Class 1 substances and set a phase-out schedule of no more than seven years.

For example, methyl bromide (ODP estimated by EPA at 0.7) was added to the list in December 1993, requiring its phaseout by January 1, 2001; this decision was altered by Congress in 1998 to harmonize the U.S. methyl bromide phase-out schedule with the 2005 deadline set by the parties to the Montreal Protocol in 1997. Also, EPA is required to add any substance that is known or may be reasonably anticipated to harm the stratosphere to the list of Class 2 substances and set a phase-out schedule of no more than ten years.

Title VI contains several implementing strategies to avoid releases of ozone-depleting chemicals to the atmosphere, including (1) for Class 1 substances used as refrigerant—lowest achievable level of use and emissions, maximum recycling, and safe disposal required by July 1, 1992; (2) for servicing or disposing refrigeration equipment containing Class 1 and 2 substances—venting banned as of July 1, 1992; (3) for motor vehicle air conditioners containing Class 1 or 2 substances—recycling required by January 1, 1992 (smaller shops by January 1, 1993); (4) sale of small containers of class 1 and 2 substances—banned within two years of enactment; and (5) nonessential products—banned within two years of enactment.

SELECTED REFERENCES

[1] U.S. Environmental Protection Agency. Office of Air Quality Planning and Standards. *Air Trends*. Research Triangle Park, NC. Compiled annually, and available at http://www.epa.gov/airtrends/.

[2] Martineau, Robert J., Jr. & Novello, David P. (eds.) (2004). *The Clean Air Act Handbook*. 2nd edition. Chicago: American Bar Association. 728 p.

Table 3. Major U.S. Code Sections of the Clean Air Act (Codified Generally as 42 U.S.C. 7401-7671)

42 U.S.C.	Section Title	Clean Air Act, as amended
Subchapter I -	Programs and Activities	
Part A -	Air Quality Emissions and Limitations	
7401	Findings, purpose	Sec. 101
7402	Cooperative activities	Sec. 102
7403	Research, investigation, training	Sec. 103
7404	Research relating to fuels and vehicles	Sec. 104
7405	Grants for air pollution planning and control programs	Sec. 105
7406	Interstate air quality agencies; program cost limitations	Sec. 106
7407	Air quality control regions	Sec. 107
7408	Air quality criteria and control techniques	Sec. 108
7409	National primary and secondary air quality standards	Sec. 109
7410	SIPs for national primary and secondary air quality standards	Sec. 110
7411	Standards of performance for new stationary sources	Sec. 111
7412	Hazardous air pollutants	Sec. 112
7413	Federal enforcement	Sec. 113
7414	Recordkeeping, inspections, monitoring, and entry	Sec. 114
7415	International air pollution	Sec. 115
7416	Retention of state authority	Sec. 116
7417	Advisory committees	Sec. 117
7418	Control of pollution from federal facilities	Sec. 118
7419	Primary nonferrous smelter orders	Sec. 119
7420	Noncompliance penalty	Sec. 120
7421	Consultation	Sec. 121
7422	Listing of certain unregulated pollutants	Sec. 122
7423	Stack heights	Sec. 123
7424	Assurance of adequacy of state plans	Sec. 124
7425	Measures to prevent economic disruption/unemployment	Sec. 125
7426	Interstate pollution abatement	Sec. 126
7427	Public notification	Sec. 127
7428	State boards	Sec. 128
7429	Solid waste combustion	Sec. 129
7430	Emission factors	Sec. 130
7431	Land use authority	Sec. 131

Table 3. (Continued)

42 U.S.C.	Section Title	Clean Air Act, as amended
Part B - Ozone Protection (repealed—new provisions related to stratospheric ozone protection are found at 42 U.S.C. 7671 et seq., below)		
Part C - Prevention of Significant Deterioration of Air Quality		
Subpart I -	Clean Air	
7470	Congressional declaration of purpose	Sec. 160
7471	Plan requirements	Sec. 161
7472	Initial classifications	Sec. 162
7473	Increments and ceilings	Sec. 163
7474	Area redesignation	Sec. 164
7475	Preconstruction requirements	Sec. 165
7476	Other pollutants	Sec. 166
7477	Enforcement	Sec. 167
7478	Period before plan approval	Sec. 168
7479	Definitions	Sec. 169
Subpart II -	Visibility Protection	
7491	Visibility protection for federal class I areas	Sec. 169A
7492	Visibility	Sec. 169B
Part D -	Plan Requirements for Nonattainment Areas	
Subpart 1 -	Nonattainment Areas in General	
7501	Definitions	Sec. 171
7502	Nonattainment plan provisions in general	Sec. 172
7503	Permit requirements	Sec. 173
7504	Planning procedures	Sec. 174
7505	Environmental Protection Agency grants	Sec. 175
7505a	Maintenance plans	Sec. 175A
7506	Limitations on certain federal assistance	Sec. 176
7506a	Interstate transport commissions	Sec. 176A
7507	New motor vehicle emission standards in nonattainment areas	Sec. 177
7508	Guidance documents	Sec. 178
7509	Sanctions and consequences of failure to attain	Sec. 179
7509a	International border areas	Sec. 179B
Subpart 2 -	Additional Provisions for Ozone Nonattainment Areas	
7511	Classifications and attainment dates	Sec. 181
7511a	Plan submissions and requirements	Sec. 182
7511b	Federal ozone measures	Sec. 183
7511c	Control of interstate ozone air pollution	Sec. 184
7511d	Enforcement for Severe and Extreme ozone nonattainment areas for failure to attain	Sec. 185

42 U.S.C.	Section Title	Clean Air Act, as amended
7511e	Transitional areas	Sec. 185A
7511f	NO_x and VOC study	Sec. 185B
Subpart 3 -	Additional Provisions for Carbon Monoxide Nonattainment Areas	
7512	Classification and attainment dates	Sec. 186
7512a	Plan submissions and requirements	Sec. 187
Subpart 4 -	Additional Provisions for Particulate Matter Nonattainment Areas	
7513	Classifications and attainment dates	Sec. 188
7513a	Plan provisions and schedules for plan submissions	Sec. 189
7513b	Issuance of RACM and BACM guidance	Sec. 190
Subpart 5 -	Additional Provisions for Areas Designated Nonattainment for Sulfur Oxides, Nitrogen Dioxide, or Lead	
7514	Plan submission deadlines	Sec. 191
7514a	Attainment dates	Sec. 192
Subpart 6 -	Savings Provisions	
7515	General savings clause	Sec. 193
Subchapter II -	Emission Standards for Moving Sources	
Part A -	Motor Vehicle Emission and Fuel Standards	
7521	Emission standards for new motor vehicles or engines	Sec. 202
7522	Prohibited acts	Sec. 203
7523	Actions to restrain violations	Sec. 204
7524	Civil penalties	Sec. 205
7525	Motor vehicle and engines testing and certification	Sec. 206
7541	Compliance by vehicles and engines in actual use	Sec. 207
7542	Information collection	Sec. 208
7543	State standards	Sec. 209
7544	State grants	Sec. 210
7545	Regulation of fuels	Sec. 211
7547	Nonroad engines and vehicles	Sec. 213
7548	Study of particulate emissions from motor vehicles	Sec. 214
7549	High altitude performance adjustments	Sec. 215
7550	Definitions	Sec. 216
7551	Study and report on fuel consumption of CAAA of 1977	Sec. 203
7552	Motor vehicle compliance program fees	Sec. 217
7553	Prohibition on production of engines requiring leaded gasoline	Sec. 218
7554	Urban bus standards	Sec. 219
Part B -	Aircraft Emissions Standards	
7571	Establishment of standards	Sec. 231

Table 3. (Continued)

42 U.S.C.	Section Title	Clean Air Act, as amended
7572	Enforcement of standards	Sec. 232
7573	State standards and controls	Sec. 233
7574	Definitions	Sec. 234
Part C -	Clean Fuel Vehicles	
7581	Definitions	Sec. 241
7582	Requirements applicable to clean-fuel vehicles	Sec. 242
7583	Standards for light-duty clean-fuel vehicles	Sec. 243
7584	Administration and enforcement as per California standards	Sec. 244
7585	Standards for heavy-duty clean-fuel vehicles	Sec. 245
7586	Centrally fueled fleets	Sec. 246
7587	Vehicle conversions	Sec. 247
7588	Federal agency fleets	Sec. 248
7589	California pilot test program	Sec. 249
7590	General provisions	Sec. 250
Subchapter III -	General Provisions	
7601	Administration	Sec. 301
7602	Definitions	Sec. 302
7603	Emergency powers	Sec. 303
7604	Citizen suits	Sec. 304
7605	Representation in litigation	Sec. 305
7606	Federal procurement	Sec. 306
7607	Administrative proceedings and judicial review	Sec. 307
7608	Mandatory licensing	Sec. 308
7609	Policy review	Sec. 309
7610	Other authority	Sec. 310
7611	Records and audits	Sec. 311
7612	Economic impact analyses	Sec. 312
7614	Labor standards	Sec. 314
7615	Separability	Sec. 315
7616	Sewage treatment plants	Sec. 316
7617	Economic impact assessment	Sec. 317
7619	Air quality monitoring	Sec. 319
7620	Standardized air quality modeling	Sec. 320
7621	Employment effects	Sec. 321
7622	Employee protection	Sec. 322
7624	Cost of vapor recovery equipment	Sec. 323
7625	Vapor recovery for small business marketers of petroleum products	Sec. 324

42 U.S.C.	Section Title	Clean Air Act, as amended
7625-1	Exemptions for certain territories	Sec. 325
7625a	Statutory construction	Sec. 326
7626	Authorization of appropriations	Sec. 327
7627	Air pollution from Outer Continental Shelf activities	Sec. 328
Subchapter IV-A	Acid Deposition Control	
7651	Findings and purposes	Sec. 401
7651a	Definitions	Sec. 402
7651b	Sulfur dioxide allowance program for existing and new units	Sec. 403
7651c	Phase I sulfur dioxide requirements	Sec. 404
7651d	Phase II sulfur dioxide requirements	Sec. 405
7651e	Allowances for states with emissions rates at or below 0.80 lbs./mmBtu	Sec. 406
7651f	Nitrogen oxides emission reduction program	Sec. 407
7651g	Permits and compliance plans	Sec. 408
7651h	Repowered sources	Sec. 409
7651i	Election for additional sources	Sec. 410
7651j	Excess emissions penalty	Sec. 411
7651k	Monitoring, reporting, and recordkeeping requirements	Sec. 412
7651l	General compliance with other provisions	Sec. 413
7651m	Enforcement	Sec. 414
7651n	Clean coal technology regulatory incentives	Sec. 415
7651o	Contingency guarantee, auctions, reserve	Sec. 416
Subchapter V -	Permits	
7661	Definitions	Sec. 501
7661a	Permit programs	Sec. 502
7661b	Permit applications	Sec. 503
7661c	Permit requirements and conditions	Sec. 504
7661d	Notification to administrator and contiguous states	Sec. 505
7661e	Other authorities	Sec. 506
7661f	Small business stationary source technical and environmental compliance assistance program	Sec. 507
Subchapter VI -	Stratospheric Ozone Protection	
7671	Definitions	Sec. 601
7671a	Listing of class I and class II substances	Sec. 602
7671b	Monitoring and reporting requirements	Sec. 603
7671c	Phase-out of production and consumption of class I substances	Sec. 604
7671d	Phase-out of production and consumption of class II substances	Sec. 605
7671e	Accelerated schedule	Sec. 606

Table 3. (Continued)

42 U.S.C.	Section Title	Clean Air Act, as amended
7671f	Exchange authority	Sec. 607
7671g	National recycling and emission reduction program	Sec. 608
7671h	Servicing of motor vehicle air conditioners	Sec. 609
7671i	Nonessential products containing chlorofluorocarbons	Sec. 610
7671j	Labeling	Sec. 611
7671k	Safe alternatives policy	Sec. 612
7671l	Federal procurement	Sec. 613
7671m	Relationship to other laws	Sec. 614
7671n	Authority of Administrator	Sec. 615
7671o	Transfers among parties to Montreal Protocol	Sec. 616
7671p	International cooperation	Sec. 617
7671q	Miscellaneous provisions	Sec. 618
[29 U.S.C. 655]	Chemical Process Safety Management	Sec. 304 of CAA of 1990
[29 U.S.C. 1662e]	Clean Air Employment Transition Assistance	Sec.1101 of CAA of 1990

Note: This table shows only the major U.S. Code sections. For more detail and to determine when a section was added, the reader should consult the official printed version of the U.S. Code.

End Notes

[1] Unlike the other NAAQS pollutants, ozone is not directly emitted, but rather is formed in the atmosphere by the interaction of volatile organic compounds (VOCs) and nitrogen oxides (NOx) in the presence of sunlight. The control of ozone is, thus, based on regulating emissions of VOCs and NOx.

[2] EPA modified the ozone standard, specified in the statute as 0.12 parts per million (ppm) averaged over a 1-hour period, to 0.08 ppm averaged over an 8-hour period, through regulations promulgated in July 1997. In April 2004, the agency promulgated an implementation rule for the new 8-hour standard. Under this rule, the 1-hour standard was revoked as of June 15, 2005, and areas that had not yet attained it were converted to new classifications depending on their 8-hour concentration of ozone. As a result of court challenges, the ramifications of this conversion to the 8-hour standard are still unfolding, but in general the former 1-hour nonattainment areas remain subject to the controls specified for their 1-hour category. New nonattainment areas that did not exceed the 1-hour standard, but do violate the 8-hour standard, in general are subject to more flexible controls under Subpart 1 (Sections 171-179B) of the act. The standard was revised again in March 2008, to 0.075 ppm averaged over 8 hours, but EPA subsequently agreed to reconsider the 2008 standard, a task it expects to complete in 2011. Nonattainment areas for the new standard would be designated following its promulgation.

[3] This section of the report was written by Linda-Jo Schierow, Specialist in Environmental Policy.

[4] The 1990 amendments specified 189 pollutants, but P.L. 102-187, enacted on December 4, 1991, deleted hydrogen sulfide from the list of toxic pollutants, leaving only 188. On December 19, 2005, EPA removed methyl ethyl ketone (MEK) from the list of toxic air pollutants. The total number of listed air toxics is now 187.

[5] This section of the report was written by Larry B. Parker, Specialist in Energy and Environmental Policy.

[6] This section of the report was written by Larry B. Parker, Specialist in Energy and Environmental Policy and James E. McCarthy, Specialist in Environmental Policy.

[7] This section of the report was written by Larry B. Parker, Specialist in Energy and Environmental Policy.

[8] This section of the report was written by Claudia Copeland, Specialist in Environmental Policy.

[9] This section of the report was written by Larry B. Parker, Specialist in Energy and Environmental Policy.

INDEX

A

B

C

I

J

K

L

S

T

U

V

W

Y